Understanding Voice over IP Security

For a complete listing of recent titles in the *Artech House Telecommunications Library*, turn to the back of this book.

Understanding Voice over IP Security

Alan B. Johnston
David M. Piscitello

ARTECH
HOUSE

BOSTON | LONDON
artechhouse.com

Library of Congress Cataloging-in-Publication Data
A catalog record for this book is available from the U.S. Library of Congress.

British Library Cataloguing in Publication Data
Johnston, Alan B.
    Understanding Voice over IP security. —(Artech House telecommunications library)
    1. Internet telephony—Security measures
    I. Title II. Piscitello, David M.
    005.8

    ISBN-10: 1-59693-050-0

Cover design by Igor Valdman

International Standard Book Number: 1-59693-050-0

10 9 8 7 6 5 4 3 2 1

Contents

Foreword

VoIP is poised to take over from the century-old public switched telephone network (PSTN). But VoIP telephony does not enjoy the same privacy as the old PSTN. This is because PSTN phone calls were based on establishing a "closed circuit" between the two parties, while VoIP phone calls send packets through the Internet, which everyone knows can be easily intercepted by anyone along the way. This naturally reduces the security of VoIP phone calls. While most PSTN phone users saw no justification for encrypting their calls, relying on the natural security of circuit-switched phone calls, even the less security-aware users are more likely to see a need to encrypt calls sent over the Internet.

The threat model for wiretapping VoIP is much more expansive than the one for wiretapping the PSTN. With the PSTN, the opportunities for wiretapping were in three main scenarios. First, someone could attach alligator clips to the phone wires near your home or office. Second, they could tap in at the switch at the phone company. This would likely be done by your own domestic government law enforcement agency, with the phone company's cooperation. Third, international long-distance lines could be intercepted by intelligence agencies of either your country, the other party's country, or a third country. It generally means a reasonably resourceful opponent.

That sounds like a lot of exposure to wiretapping, but it pales in comparison to the opportunities for interception that VoIP offers your opponents. Imagine that one of the PCs in your corporate offices becomes infected with spyware that can intercept all the IP packets it sees on the network, including VoIP packets. It could record these conversations on the hard disk as WAV files, or in a compressed audio format. It could organize them by caller or callee, allowing someone to conveniently browse through them like a TiVo player, but remotely. They might want to only listen to the CEO talking to his counterpart in another company about a merger or acquisition. Or maybe the in-house corporate counsel talking to an outside law firm. Point-and-click wiretapping, from the other side of the world. This could be done by a foreign government such as China. Or the Russian mob, some other criminal organization, or a freelance hacker. The attacker need not have the resources of a major government. He need not have the legal access that domestic law enforcement has. He can be anyone in the world, with global reach, without a military budget. In fact, sophisticated wiretapping software could even be used by countless unskilled script kiddies to wiretap their victims remotely. Wiretapping could go retail.

Think of all the criminal exploitation of the Internet going on today. Identity theft. Nigerian e-mail scams. Phishing. Millions of zombies taking over PCs everywhere, to be used as platforms to blackmail businesses with threats of distributed denial-of-service attacks. Criminal enterprises are now making more money from these activities than from selling drugs. Up to this point (January 2006), VoIP has been spared from the unwanted attentions of organized crime, because it isn't big enough to be attractive yet. Just like the Internet as a whole was some years back. When VoIP grows big enough to present lucrative opportunities for criminal exploitation, the bad guys will be all over it, like they are today with the rest of the Internet. Just think of the insider trading opportunities for someone who can wiretap anywhere in a corporate target with a mouse click. This could be more lucrative and harder to detect than identity theft.

As our economies become more globalized, even small- and medium-sized companies are opening offices in countries with cheap labor markets. VoIP becomes an enabling technology for businesses to become globalized, cheaply tying overseas offices together with the home office. Now intracompany phone calls are crossing national borders. These cheap labor markets are sometimes in countries where governments have poor track records in wiretapping their people, and too often have their law enforcement agencies influenced by organized crime. This exposes globalized businesses more and more to wiretapping by criminal enterprises and foreign governments.

While my interests tend to gravitate first to protection against eavesdropping, many other forms of attack are possible in the VoIP world, and this book attempts to address a wide range of them. So many of these attacks have never been an issue in the PSTN world, but we face them in the VoIP world. If you want to work in the VoIP industry, you'd better become familiar with them, and this book will help you understand how they work and what can be done about them. It's the first book to present both the present and future of VoIP security.

Cryptography plays an important role in protecting VoIP against many of these threats. The question of whether strong cryptography should be restricted by the government was debated all through the 1990s. This debate had the active participation of the White House, the NSA, the FBI, the courts, the Congress, the computer industry, civilian academia, and the press. This debate fully took into account the question of terrorists using strong crypto, and in fact, that was one of the core issues of the debate. Nonetheless, society's collective decision (over the FBI's objections) was that on the whole, we would be better off with strong crypto, unencumbered with government back doors. The export controls were lifted and no domestic controls were imposed. This was a well-thought-out decision, because we took the time and had such broad expert participation.

If we are to move our phone calls from the well-manicured neighborhood of the PSTN to the lawless frontier of the Internet, we'd be irresponsible not to protect them with strong encryption. We have no choice but to act. Of course, some in law enforcement might view that as an encumbrance to legitimate wire-tapping. But I see it as our duty as security professionals to protect our nation's critical infrastructure and to protect our economy, and it ought to be considered as another facet of national security.

Philip Zimmermann
March 2006

Acknowledgments

We thank our colleagues in the Internet standards community and the security practitioners struggling to secure VoIP, whose good works made this book possible. We are particularly grateful for the technical editing performed by Dave's long-time business partner, Lisa Phifer, and the candid and always insightful comments of Marcus Ranum.

1

Introduction

More voice calls are now placed over IP networks than ever before. Before too long, we would venture to say that Voice over Internet Protocol (VoIP) will be nearly commonplace. There has been a rather remarkable change in the rate of adoption. Even as recently as a year ago, the bulk of VoIP traffic was privately switched by large enterprises. Today, VoIP technology is second only to wireless LANs in consumer market adoption. As products are commoditized and today's fledgling public services mature, new voice-data applications will be offered, encouraging even broader adoption, which, in turn, will accelerate communications medium *agnosticism*. We fully expect VoIP use to quickly expand from enterprise and public broadband access networks into small business and home office LANs and wireless LANs, public and private. Mobile applications will just as surely play a major role in this expansion. Consider the popularity of today's multi-purpose cellular handheld devices, and imagine the impact broadband wireless Internet will have on mobile applications. Having a single handheld device that serves both office and personal needs for voice, messaging, computing, browsing, gaming and secure business application access, over any flavor of RF, is only a matter of time.

Voice over IP promises to be as disruptive an application as we have seen in the relatively short lifespan of the Internet. However, VoIP users and service providers must become familiar with, and develop defenses against and countermeasures for the litany of attacks that can be perpetrated against VoIP applications, devices that host these applications, and the communications services that connect them.

1.1   VoIP: A Green Field for Attackers

Users and service providers hear the term VoIP and think, "Voice-enabled applications." The attacker hears VoIP and thinks, "Here's a new phone service I can exploit, much as I have exploited telephone and cellular services in the past" or, "Here are a new set of applications and protocols I can probe for protocol specification and implementation flaws" and finally, "Here are new and untested devices I might exploit using traditional IP protocol exploits."

Why attack VoIP devices and networks? The motivations are the same for VoIP attackers as they have been for traditional telephone phreakers, cell phone snarfers, and general voice hackers: to benefit financially, via toll fraud, identity and information theft; to gain notoriety by disrupting service; and to inconvenience users and embarrass service providers. Many attacks we expect to see launched against VoIP users and networks will be similar to attacks launched against cellular and landline phone services.

Other attacks will be familiar to IT staff and network administrators because they are identical to attacks launched daily at PCs, web servers and network equipment. This is because VoIP phones and computers running VoIP software (softphones) are more computer than phone. They have operating systems and file systems and use Internet protocols. Many devices will not only handle voice calls but support messaging and data applications and self-administration as well. Generally, VoIP endpoint devices will be as vulnerable to unauthorized access, privilege escalation and system misuse, viruses and worms, and all the classic denial of service attacks that exploit network protocols as any Internet-enabled device.

Public and private VoIP service providers alike should prepare for the kinds of attacks traditional telephony service providers have witnessed against cellular and landline systems. These include toll fraud, identity and information theft, and service disruption. VoIP service providers should also prepare for attacks commonly directed at data IP networks today. Proxy servers, call managers, and, gateways are all targets for attempted unauthorized access, privilege escalation and system misuse, viruses and worms, and denial of service (DoS) attacks. Call servers can be flooded with unauthenticated call control packets. Voicemail and messaging services can be targets of message flooding attacks. Floods, subscriber impersonation, and rogue VoIP phone connections create nightmarish customer care scenarios for service providers. Resolving disputes with customers who are victims of subscriber impersonation, call hijacking, and rogue phone connections, or customers who are billed for thousands of unsolicited flood messages is a resource and revenue drain. Finally, VoIP service providers who offer online payment and customer self-administration must defend against attackers seeking to compromise accounts and databases using a variety of web attacks.

Assuming they can mitigate service disruption threats and maintain acceptable call quality, VoIP service providers must also consider the adverse effects that "security incidents" may have on user (consumer) confidence. A public VoIP phone service that makes headlines following an attack isn't likely to attract businesses or consumers. A privately operated IP PBX or voice mail server that is attacked could leave an enterprise in violation of privacy regulations.

1.2 Why VoIP Security Is Important

In discussing the need for more security in VoIP services, we have been asked why seemingly higher levels of security are required from VoIP than traditional (PSTN) telephony services. Today, our telephone conversations are fairly secure due to the perimeter and physical security of the PSTN. The nature of the Internet effectively removes both of these barriers, making eavesdropping and tapping a trivial effort in many cases. Because of this, we must add security measures to protect VoIP in the application's signaling and media protocols. Alternatively, we can use one or several of the excellent Internet security protocols commonly in use today to protect web and other data applications. We must also further consider security measures to protect VoIP infrastructures and end devices that are now commonly viewed as industry best practices.

Some argue that VoIP is not very different from electronic mail. E-mail is retrieved and sent from one computer to another with token regard for authentication, confidentiality and message accuracy. Why should VoIP be any different? The proliferation of spam and phishing attacks, not to mention the constant threat of e-mail-delivered viruses and worms, amply illustrate that e-mail is an undesirable role model for important and sensitive communications. Users are accustomed to and expect a high level of accuracy and confidentiality when they place voice calls today. VoIP ought to strive to meet or exceed user expectations and this means offering a more secure service than email as early in the adoption phase as possible.

Deploying VoIP securely is challenging but not impossible. Part of the reason why it is challenging is that, like so many Internet protocols and applications, security features for VoIP have been introduced as an afterthought rather than having been incorporated into the original design. Because it is not uncommon for new applications to be deployed without waiting for security services, we have ample experience in compensating for these circumstances. Moreover, even if security features had been available from the earliest adoption of VoIP, complementary security considerations would still be necessary, because VoIP is simply one of many Internet applications.

End-to-end security between VoIP endpoints and defense in depth (layered security) are important themes to be discussed in this book. We will discuss methods for protecting VoIP endpoints, network equipment, and individual networks. Such methods are necessary, but ultimately not sufficient. A complex set of protocols such as VoIP requires a true end-to-end security architecture and correspondingly secure protocols. Why? Consider two users, Alice and Bob, who communicate over VoIP. Alice may implement good security at her VoIP endpoint and network, use a reputable VoIP service provider, and employ encryption over her wireless LAN. Bob, however, may use a shared Ethernet segment. IP traffic transmitted over his LAN, including VoIP traffic is visible to other devices, and, much like a party line or phone tap, other parties can listen in on his conversations. Only if Alice and Bob employ end-to-end security will their VoIP calls provide confidentiality.

1.3   The Audience for This Book

This book is written for moderately technical individuals and professionals who are interested in using Voice over IP for business and personal use, or, those persons who are employed by organizations to deploy and operate VoIP networks and applications. Readers are assumed to have a basic knowledge of the Internet, Internet security, VoIP concepts and protocols, such as Session Initiation Protocol (SIP). While a complete introduction to all these concepts and protocols is not provided, we do attempt to introduce the basics of security as they apply to Internet communication in general, and VoIP specifically.

We consider VoIP security from architectural, design and high-level deployment points of view. However, we will not attempt to provide a list of known vulnerabilities in commercial products. Instead, this book should enable an engineer or manager to appreciate the issues associated with securing VoIP. It may provide you with useful questions for your vendor or service provider to determine how secure their product or service is. If you are designing a service, we hope it will provide you with some current and future directions in securing VoIP.

1.4   Organization

Chapter 2 provides an introduction to cryptography. Cryptography plays an essential role in securing VoIP applications and services. We explain basic cryptographic concepts as encryption and decryption, authentication, and integrity protection in nonmathematical terms.

Chapter 3 introduces Voice over IP architecture and concepts. We provide a high-level description of the relevant VoIP protocol standards of the Internet Engineering Task Force (IETF), including Session Initiation Protocol (SIP),

Session Description Protocol (SDP), and Real-time Transport Protocol (RTP). We also briefly discuss International Telecommunications Union standard H.323, and certain proprietary VoIP protocols.

Chapter 4 introduces the threats against VoIP users and service providers, and describes the types of attack that users and service providers can expect to encounter. While not an exhaustive list of every known vulnerability and exploit, the threats and attacks described here provide useful insight into the vectors and methodologies that will most commonly be used to attack VoIP endpoints, network equipment and networks.

Chapter 5 provides readers who are unfamiliar with Internet security terminology with a good foundation. Chapter 6 introduces Internet security protocols that are used to provide authentication, confidentiality and integrity services for data and voice applications. We discuss the principles of virtual private networking and tunneling, and provide high-level explanations of how such protocols as IP Security (IPSec), Transport Layer Security (TLS, also known as Secure Sockets Layer protocol, SSL), and Secure Shell (SSH) can be used to protect voice signaling and media connections.

Chapter 7 discusses client and server security principles and concepts which are commonly applied to data network endpoints and network equipment, and which also should be considered for securing VoIP systems.

Chapter 8 considers the security concept of authentication and the roles the different kinds of authentication play in VoIP applications. Chapter 8 also discusses the importance of auditing and accounting in voice networks in general, and VoIP networks in particular.

Chapters 9 through 13 describe end-to-end security for VoIP. Those with a high level of knowledge of VoIP and Internet security principles and concepts may want to access these chapters immediately to get the key concepts. Chapter 9 discusses how to secure VoIP signaling using techniques such as HTTP Digest, S/MIME, and TLS. Chapter 10 discusses securing VoIP media flows, discussing new protocols such as Secure RTP (SRTP) and Multimedia Internet Keying (MIKEY). We also discuss the principles of key management and agreement. Chapter 11 discusses identity and the ways in which identity can be asserted and verified in a VoIP network. The use of self signed certificates, and the Secure Assertion Markup Language (SAML) are covered. Chapter 12 discusses Public Switched Telephone Network (PSTN) security and gateway security. Here we discuss some of the issues specific to PSTN interconnection, including Gateway security, PSTN (toll) fraud, and telephone number mapping in the DNS. Chapter 13 describes a potential problem of spam or unsolicited, bulk VoIP calling. Chapter 14 contains conclusions and summaries of the main points.

2

Basic Security Concepts: Cryptography

2.1 Introduction

This chapter introduces the basic cryptographic concepts including authentication, confidentiality, and integrity protection. The treatment of these topics is introductory and sufficient for understanding how cryptography is applied in VoIP security measures. For an in-depth look at these topics, refer to [1].

2.2 Cryptography Fundamentals

The problems cryptography attempts to solve are nearly as old as humankind, and can be summarized in the following example. Dave has some private correspondence to send to Alan and wants some assurances that the message will be delivered in such a fashion that only Alan can access it. Specifically, Dave wants assurance that no other party delivering the message or any party that might intercept the message can either read or alter it.

As the recipient of this private correspondence, Alan has his own set of needs. Alan wants to be as certain as possible that the message he's received really came from Dave, and not an impostor. Alan also wants some assurance that the message wasn't delayed in delivery, rendering it obsolete, or possibly indicating that some party withheld it from delivery long enough to read or modify it.

This example illustrates the fundamental security services cryptography attempts to provide:

- Authentication assures the recipient (Alan) that the sender (Dave) is who he claims to be. Authentication is the method by which a user or endpoint proves his (its) identity to another. Authentication can also be the basis upon which a service or provider determines whether a user can be permitted or *authorized* to use a service or resource.

- Integrity assures Alan that the message Dave sent has been delivered without alteration, and within a reasonable timeframe. Integrity protects against both malicious attacks and communication errors and failures.

- Confidentiality assures Dave that only Alan can read the message. Confidentiality prevents parties other than Alan from reading the message. It does not necessarily hide the fact that Dave and Alan are corresponding, however.

- Non-repudiation assures Alan that Dave cannot deny he created the message. Non-repudiation refers to an authentication that cannot be refuted. Non-repudiation provides verifiable proof in other situations as well, namely:

  An individual responsible for an approval process cannot deny having approved an action, or a message's content.

  A delivery authority cannot deny having accepted a message for delivery, or having delivered it to the recipient.

  A recipient cannot deny having received a message.

Let's apply the example to a typical communication over the Internet using e-mail. Dave and Alan exchange mail messages in plain text that is easily readable by Dave, Alan, and all the systems that provide email delivery between them. The message is also easily observed while it is transmitted over popular communications media such as shared segment Ethernet LANs and Wireless LANs. In plain text format, any man in the middle (MitM) who captures the message during its transmission can easily read email messages and headers. This same MitM can pose as the sender (Dave) or the recipient (Alan). A MitM can alter the message itself, modify headers, and deliver the message to parties other than Dave and Alan.

If Dave and Alan really want to keep their e-mail correspondence private, they must convert at least the portion of the message they wish to keep confidential into something that's not easily readable by anyone but Dave and Alan. Specifically, Dave wants to convert plain text to something that only Alan can read; in cryptographic terms, Dave wants to encipher or encrypt the plain text correspondence into cipher text before he sends it to Alan. Moreover, he wants to perform this encryption in such a manner that only Alan can recover or decrypt the original plain text of the message from the cipher text he receives. This is the 100,000-foot view of encryption.

Bruce Schneier, author of *Applied Cryptography*, widely acknowledged as the bible of cryptography, illustrates the mathematical relationship between plain text (P), cipher text (C), and the encryption (E) and decryption (F) functions – and the important recovery identity as follows:

```
E(P) = C        Original function of encryption
D(C) = P        Reverse process of decryption
D(E(P)) = P     Recovery of original plain text from
                cipher text
```

Graphically, Schneier represents the process as depicted in Figure 2.1.

Seeing the relationship illustrated in this manner, it is clear that Dave and Alan must agree on some function or process they will employ to take plain text and convert it to cipher text in a reversible way. This function is called a cryptographic algorithm.

As the name implies, key-based algorithms rely on a key. Rick Smith, in *Internet Cryptography*, describes a key as "a special piece of data — e.g., a large, randomly chosen number — that directs the cryptographic device to encrypt [or decrypt] a message in a distinctive way" [2]. Keeping keys secret is more manageable than keeping algorithms secret or restricted (see [1], page 11).

Modern cryptography commonly uses publicly reviewed and published key-based algorithms. These are widely accepted for use to secure Internet applications because they have been subjected to considerable scrutiny or cryptanalysis and have been demonstrated to be provably hard to break without the disclosure of the encryption (decryption) keys, so their quality is less subject to question than a proprietary cryptographic algorithm.

The strength of the encryption, entropy, is related to the length of the encryption key. For a given cryptographic algorithm, the longer the key, the more secure the encryption. This can be readily appreciated by understanding the methods, which can be used to break encryption. For example, an attacker can randomly try all possible key combinations until the resulting plain text is revealed. The time and processing required to "brute-force" the plain text from cipher text with zero knowledge of the plain text is affected by the number of possible combinations, which is directly related to the number of bits. Some knowledge of the plain text (e.g., protocol header information that contains

Figure 2.1   Relationship of plain text to cipher text.

fixed or known values) can greatly speed up this approach and is known as "crib-bing." However, knowledge of the plain text provides little assistance when well designed and industry accepted encryption ciphers are used with appropriately long keys.

Two kinds of key-based cryptographic algorithms are commonly used by Internet security applications: symmetric or secret key cryptography and asymmetric or public key cryptography.

2.2.1   Secret Key (Symmetric) Cryptography

Secret key cryptography uses a single key for both encryption and decryption operations. The property of symmetry is a fundamental property of this form of cryptography, that is, both parties have the same key. The important conditions for successful application of symmetric cryptography are that (a) the parties exchange or share knowledge of a common key before they use it, and (b) the value of the key is kept secret among the parties. These properties lend this form of cryptography a second name, shared secret key. Shared secret key cryptography is often incorrectly called private key cryptography, because the words private and secret are mistakenly interpreted as interchangeable. A private key is an entirely different kind of key, specific to public-private key cryptography, which we describe later.

To illustrate shared secret key cryptography, consider Dave's desire to secure his email correspondence with Alan. Suppose Dave encrypts and sends messages to Alan using a well-known symmetric cryptographic algorithm and a key he and Alan have shared prior to any encryption (in Figure 2.2, the hexadecimal string 0xae014f3001). As long as Dave and Alan keep this key secret, only Alan (and Dave) can decrypt Dave's correspondence; many people may know the algorithm, but only Alan shares knowledge of the secret key with Dave. Similarly, Alan can encrypt and send a reply to Dave's correspondence using the same algorithm and key.

Schneier and others depict the operation of shared secret key cryptography as we depict in Figure 2.2.

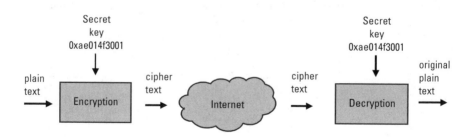

Figure 2.2   Shared secret key cryptography.

2.2.1.1 Shared Key Encryption Algorithms

There are two main types of symmetric key encryption algorithms: block ciphers and stream ciphers. Block ciphers encrypt fixed length units of plain text data known as blocks. A block cipher operates with the key on a single block of plain text. A stream cipher, on the other hand, operates with the key on a block of plain text and the internal state of the algorithm, which depends on previous blocks of plain text.

The Data Encryption Standard (DES) [3] is an example of block cipher. DES utilizes a 64- bit key, but 8 bits are parity bits, making the effective key length 56 bits. DES operates on a block of length 64 bits. DES is no longer considered as resistant to attack as it once was. Significant advances in computing power since the publication of DES make it relatively easy to try all possible key values in a reasonable time period. Several applications are able to reveal DES keys, and some are able to break keys in less than 24 hours of computation. Today, the technique of using multiple (iterative) applications of DES encryption using multiple keys is applied to make the DES-based encryption more resilient against attacks (and extend the lifetime of chip sets, software, and hardware that compute the DES algorithm). Triple DES encryption can be performed using DES iteratively using two 56-bit keys and three independent 56-bit keys. Both thwart MitM attacks. For example, using three independent keys, the encryption of ciphertext (C) begins by encrypting plain text (P) with key #1 ($k1$), then decrypting the ciphertext output of the first computation with key #2 ($k2$), and finally encrypting the output of the second computation with a third key ($k3$), or $C = (E_{K3}(D_{K2}(E_{K1}(P))))$. The corresponding decryption to plain text (P) = $(D_{K3}(E_{K2}(D_{K1}(C))))$. By employing an "encrypt-decrypt-encrypt" (EDE) sequence, compatibility with DES implementations is achieved (e.g., when $k1 = k2 = k3$ the 3DES implementation represents a single-key DES implementation) [4].

The IDEA (International Data Encryption Algorithm) [5] block cipher uses a 128-bit key to encrypt blocks of 64 bits. IDEA performs a "mix" of algebraic operations on 16-bit sub-blocks of the plaintext block using 16-bit sub-keys of the 128-bit key (see [1], p. 320 for an easy to follow description of the algorithm). The algebraic operations are easily implemented in hardware and software, and encryption processing is faster than DES. IDEA was used in version 2.0 of Pretty Good Privacy (PGP), which is described in Chapter 6.

Another modern encryption cipher is known as AES (Advanced Encryption Standard) [6]. AES was developed as the successor to DES, and is a block cipher with a block length of 160 bits and keys of length 128, 192, and 256 bits. AES is commonly used in cipherblock chaining (CBC) mode. CBC mode utilizes feedback that effectively turns a block cipher such as AES into a stream cipher. The CBC algorithm works as follows: the encryption of the current block depends on the encryption key, the current plaintext block, *and* the previous n ciphertext blocks. With this feedback, the algorithm needs to be initialized with an initialization vector (IV). This is shown in Figure 2.3 below.

Many other symmetric encryption algorithms are used on the Internet today. Rivest Cipher #4 (RC4, [7]) is a proprietary stream cipher developed by RSA Security, Inc. and is widely used in Internet security protocols. RC4 is used in wireless equivalent privacy (WEP) and is one of several encryption algorithms that can be negotiated for use with TLS (see Chapter 6).

2.2.2   Asymmetric (Public Key) Cryptography

Symmetric encryption is very powerful but has some obvious limitations. If Dave and Alan want to exchange encrypted e-mail with other parties, they have to either share their secret key among a group, or keep a separate shared secret with whomever they intend to exchange encrypted e-mail. Group shared secrets are easy to maintain: one secret per group is more manageable than one per e-mail correspondent. However, the more people who know a secret, the harder it is to keep. Moreover, while a secret key shared between two parties has some non-repudiation properties, a secret key shared among dozens or hundreds for all practical purposes has none. Conversely, maintaining unique shared secrets

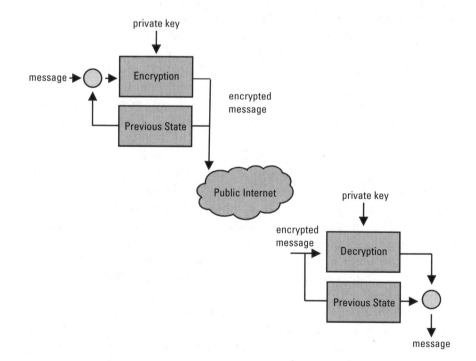

Figure 2.3   Cipherblock chaining (CBC) mode block encryption.

for everyone you wish to e-mail using encryption quickly becomes unmanageable. Moreover, with either alternative, it is administratively difficult to change keys with reasonable frequency.

To solve these problems, we look to asymmetric encryption. Asymmetric encryption uses a pair of keys that have a unique mathematical relationship: if you encrypt a message with one key *A* of the key pair *AB*, you can only decrypt with the other key, *B*, and vice-versa. Schneier, Perlman [8], and others depict the operation of asymmetric key cryptography as shown in Figure 2.4.

In practice, to achieve the goal of message confidentiality, we must make it possible for many people to encrypt a message, but to assure that only the intended recipient is capable of decrypting the cipher text. Since this goal is accomplished when an individual keeps his decryption key private and makes his encryption key generally or publicly known, asymmetric cryptography is also known as public key cryptography. Using public key cryptography, Dave can publish his public key in readily accessible repositories such as directories and key distribution servers. Anyone can encrypt a message and send it to Dave using Dave's public key, with confidence that so long as Dave hasn't disclosed his private key (or had it compromised or stolen), only Dave will be able to decrypt the cipher text of that message. Public key cryptography thus has measurably better scalability than secret key cryptography. And since private keys are never shared by individuals (or used by multiple computers), a private key is a substantially better piece of information on which to base authentication. One drawback to public key cryptography, however, is that, relative to secret key encryption, it is computationally more intense, so encryption of large amounts of data is generally slower.

2.2.3   Integrity Protection

Integrity protection provides a way for a recipient of a message to verify that a message has been delivered without modification, that is, the message delivered to the recipient is an exact copy of the message composed and sent by the originator. Integrity protects against both malicious attacks and communication errors and failures.

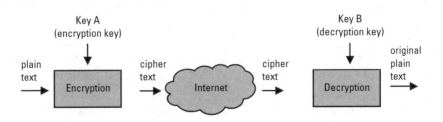

Figure 2.4   Operation of asymmetric key cryptography.

Many data communications protocols use a simple checksum algorithm as a mechanism for integrity protection. Each byte of data in the datagram or packet is summed as it is sent. A certain number of bits of this checksum (modulo n) are then appended to the datagram or packet. The receiver performs the same addition of bytes and compares the computed checksum to the received checksum. Note that if the checksum does not have any integrity protection itself, this method only protects against communication channel errors, not malicious attacks, as an attacker with the ability to change the contents of the datagram would also change to contents of the checksum to match.

2.2.3.1 Message Digests

A message digest or hash algorithm is another simple method of providing integrity protection. A message digest is a one-way mathematical function performed on an arbitrary length set of data (the message) to produces a fixed length output (the digest, or hash). The algorithm is chosen to minimize the possibility that two different messages will produce the same message digest (known as a collision). The probability that any two strings of data submitted to a hash function will yield the exact same hash value is so small that cryptanalysts consider hash values to be practically unique to the string of data from which they were derived. Thus, a hash can be thought of as a mathematical fingerprint of the original message. As we will show in Section 2.2.3.3, this property is the basis for proving the integrity of a message because for a given hash value, an eavesdropper can't capture the hash value, and easily create and substitute another message that has the same value.

Two additional properties make hash values useful in providing message integrity. Hash values are small, so they can be appended to a message without undue overhead. They are computed very quickly, so message integrity can be added without significant processing overhead and latency.

Given these properties, we can begin to explain how message integrity is provided in communications. Dave wants some assurance that a message he has received from Alan hasn't been tampered with by the nefarious "man in the middle." To accommodate Dave's concern, Alan computes a hash value on the message he intends to send and appends this value to the message before sending to Dave. It seems that no one who intercepts and attempts to alter this message can generate the same hash value from the altered message as Alan generated for the original message. Dave can thus confirm the message he's received is the same as the original sent by Alan by computing his own copy of the hash value, using the same hash algorithm as Alan, and comparing his result to the hash attached to Alan's message.

However, appending a message digest and sending this along with a message in this manner leaves the message vulnerable to a substitution attack. Unless some protection is added to the hash, a MitM can alter the message, compute a hash value on the altered message, and forward the altered message and digest to the recipient, who has no means of detecting the change. When message digests are used in this manner, message authentication must be used (see Section 2.2.3.3).

2.2.3.2 Digest Algorithms

The simple checksum is an example of a message digest. Digest algorithms which are more resistant to attacks include the message digest algorithm (MD-5 [9]) and secure hash algorithm (SHA-1 [10]). The MD-5 hash produces a 128-bit digest while SHA-1 produces a 160-bit digest. MD-5 is the simpler of the two to compute, but SHA-1 is considered stronger. The testing and exploration of digest functions remains an ongoing project in the security community.

2.2.3.3 Message Authentication and Digital Signatures

Message hashes prove a message hasn't been altered. But how can Dave be certain that the message he's received is truly from Alan? Suppose the message is not from Alan: how could Alan repudiate a claim that it is? Secure communications should be able to demonstrate authenticity (provide non-repudiation) of origin. We can combine message hashing and public key encryption to provide message authentication and integrity in the following way.

Alan and Dave agree to use public-private key pairs and make the public keys available. Alan sends a message to Dave. For message integrity, Alan computes a hash value over the message. Then Alan encrypts the message hash value using his private key. This produces what is known as a message authentication code (or digital signature). Alan now sends the message and the digital signature to Dave.

Dave receives the message and the appended digital signature. To verify the integrity of the message, Dave must first decrypt the hash value from the digital signature Alan appended. The hash is encrypted with Alan's private key so Dave obtains Alan's public key and uses this to decrypt the digital signature. Now Dave has the hash value Alan computed and appended. He computes the hash of the message he receives, and compares this to the hash value Alan signed and appended to the message. If the two hash values are the same, he can conclude that:

- The message he has in his possession is exactly the same as the one Alan sent: no one has modified the message in transit.
- Alan signed the message using his private key. Dave knows this because he could only have decrypted the digital signature using Alan's public key if this were true.

- Alan can't repudiate Dave's claim that Alan sent this message since only Alan's private key could have created the digital signature appended to the message.

Alan and Dave have provided measures for message authentication, integrity and non-repudiation. Note that in this example, message authentication and integrity protection are complementary. Without using message authentication, a MitM could alter the message. Without integrity protection, a MitM could attempt to delete the digital signature and forward the message without the authentication code. Also note that in this discussion, Alan and Dave are not encrypting the message itself. An eavesdropper could be passively monitoring and recording messages Alan sends to Dave, so they still need to provide some additional measures—message confidentiality—if private correspondence is required.

Digital signature algorithms in use today include the RSA [11] algorithm, which uses identical algorithms to perform asymmetric encryption and decryption. The United States government's Digital Signature Standard (DSS)[12], which specifies a digital signature algorithm (DSA). This algorithm is only useful for computing signatures and cannot be used for asymmetric encryption and decryption. DSS utilized the SHA-1 message digest hash.

The hashed message authentication code (HMAC) [13] algorithm combines a message digest hashing function with a shared secret encryption key to provide integrity protection and authentication. When used with SHA-1, the algorithm is known as HMAC-SHA-1. The encryption key is chosen to be the same length as the message digest hash. Figure 2.5 shows the operation of the HMAC algorithm.

The hash function provides integrity protection. If Dave is able to decrypt the encrypted hash, he can conclude that the encrypting party knows the shared secret key, and if neither Dave nor Alan revealed this shared secret, Dave knows that Alan sent the message. Thus, processing of the HMAC provides a message authentication.

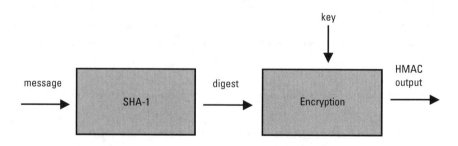

Figure 2.5 Hashed message authentication code algorithm HMAC-SHA-1.

2.2.4   Authenticated and Secure Key Exchange

We now know how to apply cryptography to provide confidentiality and that is to encrypt the message. Now we consider how parties acquire or derive encryption keys.

Earlier, we explained that secret key cryptography is fast, but distribution of the shared secret keys is difficult. While public key cryptography offers a good solution for key distribution, encryption using public keys isn't fast enough to use on large messages. As messages become longer, the difference between secret and public key encryption and decryption becomes more pronounced.

To counter this, Dave and Alan want to use shared secret key cryptography for message encryption. In order to do so, they need to solve the following problems:

> Dave and Alan need to establish a shared secret key before it can be used in symmetry cryptography. How can they exchange keys in private if they don't know the secret key(s) beforehand?

> If Dave and Alan *do* figure out a way to exchange keys, how can Alan be certain he's really exchanging a secret key with Dave?

> How can Dave be certain he's exchanging a secret key with Alan?

The properties of asymmetric (public key) cryptography suggest part of the solution: perhaps we can use public key cryptography to securely distribute symmetric keys? Suppose Dave creates a secret key and encrypts the key with Alan's public key before sending it to Alan. Dave can be certain that only Alan will be able to decrypt the secret key (using Alan's private key). Suppose Dave also signs the encrypted secret key with his own private key. Alan will know that only Dave could have sent the key.

Two popular methods that apply this hybrid cryptographic strategy to protect key exchange and add confidentiality are the Rivest, Shamir, and Adleman, RSA [14] and Diffie-Hellman [15, 16] cryptographic algorithms.

2.2.4.1   Adding Protected Key Exchange and Confidentiality Using RSA

Let's assume that Dave wants to send Alan a private (cryptographically protected) message. One way he can do this is to encrypt the message with a (large) random number as the key for a symmetric cryptographic algorithm. Pseudo-random number generators (PRNGs) can produce values that are sufficiently random for this purpose.

Dave has to somehow get this key to Alan so Alan can decrypt the message. Here's what they agree to do.

Dave and Alan generate RSA key pairs. They make the public keys of these pairs known to each other. Dave takes the large random number he uses to encrypt the message, and encrypts it using the RSA algorithm and Alan's public key. Dave adds integrity protection by hashing the message using an MD-5 or SHA-1 hash algorithm. Dave encrypts the hash value using his private key. Dave appends this digital signature to the message and sends the message to Alan.

Dave has now created a message with these properties and assurances for Alan:

> The message and encryption key are undeniably from Dave. Alan can corroborate this claim by processing Dave's digital signature.

> Only Alan can decrypt the secret key (the large random number Dave generated). It is encrypted with Alan's public key.

> The secret key Alan now shares with Dave is the only one that will decrypt this message.

Alan now decrypts the message, computes a hash value from the message he decrypts, and compares this to the hash Dave appended to the message and signed. If the hash values are equal, the message is unaltered.

The exchange described here is used to encrypt single messages and would be practical for exchanging confidential e-mail messages and other applications where it is convenient to send the secret key along with the encrypted message. It is also a convenient algorithm for peer authentication. However, many Internet security protocols must exchange secret keys before endpoints begin encrypting all the traffic they exchange. For forms of secure communication where large volumes of messages are encrypted over possibly long time periods, many security protocols use the Diffie-Hellman public key exchange.

2.2.4.2   Diffie-Hellman Public Key Exchange

Diffie-Hellman is a public key exchange method that has two remarkable properties:

> Two parties can generate the same secret key, independently.

> The secret key is never transmitted from one party to the other.

Diffie-Hellman (DH) is a very secure way to generate dynamic, short-lived session keys over an otherwise un-secured connection. The DH algorithm was published in 1976 and has since found widespread use in encryption systems on the Internet. The algorithm works as follows, with the steps listed below, and is illustrated in Figure 2.6.

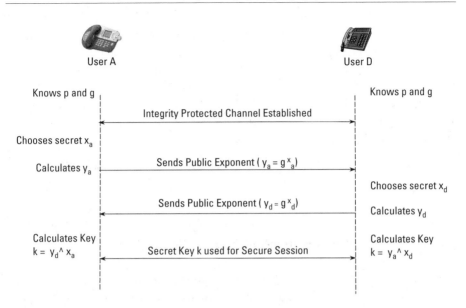

Figure 2.6 Diffie-Hellman Key Exchange Protocol.

Alan and Dave start with a large prime number p, and a generator g, which generates a set of prime numbers from 0 to p-2.

Alan initiates the DH exchange by choosing a private exponent, x_a, which is a member of the generator set. Alan calculates the corresponding public exponent ya using: $y_a = g^{\wedge}x_a$ and sends the value y_a to Dave.

When Dave receives Alan's public exponent, he randomly selects his own private exponent, x_d, also a member of the same generator set g. Dave calculates the corresponding public exponent y_d using $y_d = g^{\wedge}x_d$ and replies sending this valued back to Alan.

Alan and Dave then compute the same secret key using the formulae $k = y_d^{\wedge}x_a = g^{\wedge} x_a x_d$ modulo p and $k = y_a^{\wedge}x_d = g^{\wedge} x_d x_a$ modulo p which will yield the exact same value.

Alan and Dave now can use this secret key k for symmetric key encryption.

A DH key exchange does not require a confidential channel. The two private exponents are never sent over the channel but are generated at the time of the exchange then discarded.

A key generated using DH has perfect forward secrecy. No long-term secret is used to protect the keying material. Should the secret key be compromised, it would only affect the one session for which the key was generated. It is thus possible to minimize the exposure to key disclosure by frequently re-generating secret keys using DH.

A man-in-the-middle attacker who knows p and the generator g could sit in the middle and can participate in the DH exchange with Alan and Dave separately. Absent any authentication exchange to prevent this attack, neither Alan nor Dave can know that they have performed the DH exchange with the attacker and not with each other. As a result, the attacker could impersonate both parties and decrypt all the data from both Alan and Dave. This attack is shown in Figure 2.7.

Alan and Dave must use authentication and integrity protection to prevent this form of attack. This is often achieved by employing digital signatures (see Section 2.2.3.3)

2.3 Digital Certificates and Public Key Infrastructures

A digital certificate is used to verify that a message comes from the person (or entity) who claims to have originated it. A digital certificate contains the name and the public key of an entity, and information about that entity that distinguishes the entity from all others. Digital certificates allow us to associate IP addresses, domain names, and e-mail addresses with the public keys their owners have assigned to them, and thus we say that a digital certificate binds an entity to its public key. A digital certificate also confirms the identity of an entity, such as an e-merchant's credit card transaction server. Something more than the digital certificate is needed for us to conclude whether a certificate is legitimate or bogus. In the world of digital certificates, this something is a concept called certificate signing.

An entity that wants others to trust that its certificate is valid, and that the public key contained therein is also valid, will have some recognized trusted authority digitally sign the certificate that contains the entity's public key. Since we place such a premium on authentication, it follows that we need to create and acknowledge a trusted third party to issue and sign certificates, called a

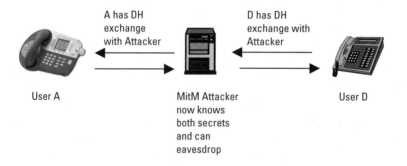

Figure 2.7 Diffie-Hellman man-in-the-middle attack.

certificate authority (CA). When a CA issues a digital certificate, it offers the certificate as proof that an entity's credentials have been vetted by the CA; that the CA has taken appropriate measures to assure that the credentials are accurate; and that from this evidence, the CA believes the entity has proven it is who it claims to be.

The important property a CA brings to authentication is that two parties who have not previously established a trust relationship can implicitly trust each other because they each share a relationship with a common and trusted third party that vouches for the legitimate identity of both parties. This third-party trust is a fundamental requirement for large-scale deployment of any system based on public-key cryptography. A registration authority (RA) authenticates and validates each party prior to a CA issuing a certificate to the party. The rules and steps associated with this authentication and validation are described in a standard document known as a certification practice statement (CPS).

A CA guarantees the authenticity and integrity of digital certificates it issues by affixing a digital signature created by the CA's signing, or private key. This private key is the critical piece of digital information that keeps any public key infrastructure trustworthy, because users verify the signature on every certificate by using the issuing CA's verification or public key. CAs must carefully protect their signing private keys against theft or accidental loss.

A CA must have a process for revoking a digital certificate whose private key has been compromised or one that no longer represents a legitimate binding between the holder and certificate (e.g., an employee that leaves a company). Because certificates are used for authentication, certificate revocation must be accomplished in real time to minimize the likelihood of misuse and unauthorized access. CAs must provide timely revocation procedures and a method for distributing certification revocation lists (CRLs), and entities that use certificates must have procedures in place for checking CRLs with adequate frequency.

We've described how effective use of digital certificates and public keys requires administrative measures necessary to make the public key, its associated attributes, and status widely available in an automated, reliable, and interoperable manner. A public key infrastructure [18–21] is needed to perform key and certificate creation, distribution, lookup, backup, recovery, and revocation services for a community that uses digital certificates. These services are very similar to those performed by a passport-issuing authority. A passport authority creates secure, authorized, printed documents. Passports are an accepted form of identity verification almost anywhere in the world. Passport authorities in all countries collect and verify information on a citizen, such as name, photograph, date of birth, residence address, and the like. If the information a citizen provides satisfies its registration criteria, the passport authority will issue a passport, a document that is difficult to alter without detection. The authority imposes a lifetime on the passport, and then imprints its seal on that document, effectively affixing its unique signature in a way that can't be repudiated. Countries acknowledge passport seals of sovereign states with which they maintain diplomatic relationships. Individuals with passports rely on these documents to prove their identities to merchants and authorities while traveling from country to country.

In the digital world, independent third parties—public CAs such as Verisign, CommerceLock, and GeoTrust (formerly EquiFax)—verify credentials and issue digital certificates so that merchants can offer secure transactions. Public CAs are acknowledged as trusted parties responsible and accountable for verifying the authenticity of users or systems to whom they issue certificates, similar to passport authorities. They issue digital certificates using software from EnTrust, Baltimore Technologies, Microsoft, to name a few.

There is no single, universal public key infrastructure yet, if ever. While sovereign states can claim exclusive rights to issue passports for their citizens, there is considerable resistance to entrusting governments with encryption key administration. Thus, individual CAs, both public and privately operated, may exercise authority to issue certificates over a given user constituency. Many certificate authorities exist and will continue to exist. Cooperation and the important related concept of asserting trust between CAs remain two of the more difficult problems to resolve in the Internet today.

2.3.1 Certificate Assertions

Digital certificates commonly used today utilize a format known as X.509 [22] Certificates are encoded using ASN.1. An example certificate is shown in Figure 2.8.

An X.509 certificate is used to assert a distinguished name (DN). In this example certificate, the first field is the serial number. If a certificate is revoked and reissued, it may have the same DN but will have a different serial number. The certificate revocation list (CRL) lists revoked certificates using the serial number. The next field indicates the algorithm used for the CA signature encryption and decryption. This signature is the last element in the certificate beginning after the signature algorithm field.

The next field indicates the issuer of the certificate and can contain geopolitical information such as country, state, and city. The validity field indicates the duration for which the certificate is valid. The subject field contains the DN of the certificate. The common name or CN field identifies the certificate owner. While the CN can be populated with a fully qualified domain name of the device, the subject alternative name (subjectAltName) is a better place to put this. The subjectAltName can contains at least one of the dNSName containing the fqdn, iPaddress, rfc822Name, or uRI.

The next field contains the public key of the owner and indicates the algorithm used for encryption and decryption of its communications. The signature

```
Serial Number:
41:81:D4:94:99:DA:BB:2A:03:79:52:29:B7:A3:71:09

Certificate Signature Algorithm:
PKCS #1 SHA-1 With RSA Encryption

Issuer:
OU = www.verisign.com/CPS Incorp.by Ref. LIABILITY LTD.(c)97 VeriSign
OU = VeriSign International Server CA - Class 3
OU = VeriSign, Inc.
O = VeriSign Trust Network

Validity
Not Before:
1/22/2004 18:00:00 PM
(1/23/2004 0:00:00 AM GMT)
Not After:
2/7/2006 17:59:59 PM
(2/7/2006 23:59:59 PM GMT)

Subject:
CN = www.artechhouse.com
OU = Terms of use at www.verisign.com/RPA (c)01
OU = HORIZON1            .
O = Horizon House Publications
L = Norwood
ST = Massachusetts
C = US

Subject Public Key:
PKCS #1 RSA Encryption

Subject's Public Key:
30 48 02 41 00 9c ab f0 02 76 90 24 43 12 e8 9e
2f 8c e1 6c f6 26 98 3e 9b 06 01 74 60 95 be 16
73 5d 1e 2f 03 0f af df 38 69 b4 ca c1 d8 02 48
26 a8 23 b2 18 68 11 0c 77 b4 86 c0 1c 06 00 a3
9b f4 da 4c e7 02 03 01 00 01

Certificate Signature Algorithm:
PKCS #1 SHA-1 With RSA Encryption


Certificate Signature Value:
87 c5 c4 fd f0 68 0e 12 01 5a 21 a7 ec e9 f7 73
7d 4c 0d 59 c9 57 8a c7 5d 5f a5 b8 7e bf a4 10
8e 30 38 a0 9d 92 cc f9 09 cd d5 92 19 d7 1a c1
70 8f d5 d3 a4 cc 14 75 d2 ba fa a9 05 ee 47 e4
42 2f 68 03 1d 7d ae 99 7c 79 7e 24 28 97 6f 8c
b1 08 69 5b ec 2a 6f fc 02 09 18 6f 9a 6b 26 ee
15 59 92 c0 5b fe a0 b7 28 36 bf da ab 2a 79 7c
0d 08 71 63 14 1f 91 cb aa db 7a 0f 3e 7f f0 42
```

Figure 2.8   Example X.509 certificate.

in the next field is the certification by the CA that this is in fact the valid public key for this owner.

If the CA is in its trust hierarchy, a recipient of a certificate can validate the signature of the CA, check the serial number against a CRL, and then have a good assurance that the public key in the certificate can be used to securely communicate with the certificate owner.

CRLs are in general not very practical to use, especially since, in theory, it must be checked each time before a certificate is used. Instead, OCSP (Online Certificate Status Protocol) [23] has been proposed to provide a query/response mechanism for validating certificates in real time at time of use. An OCSP client queries an OCSP responder and receives a response of good, revoked, or unknown. An OCSP response is signed by the CA which issued the certificate in question, or by an authorized OCSP responder. Note that to properly validate a certificate, the entire certificate chain must be validated. (OCSP is supported by the Firefox browser, among others.) The PKIX Public Key Infrastructure working group [24] has recently developed SCVP (Standard Certificate Validation Protocol) [25]. SCVP allows a client to request a server validate that a public key belongs to a particular identity and that the certificate is valid for the intended use. Another use of SCVP is for a client to request a server determine the certificate chain for validation, and return this certificate chain information to the client which then performs its own validation.

The IETF SIP working group has also proposed a method of making so-called self-signed certificates useful. A self-signed certificate is one that is asserted and signed by the owner of the certificate. Since no chain of trust is invoked in this model, the certificate must be retrieved in a secure manner from a reliable source. For example, a domain may have a single standard certificate, issued by a trusted CA. Individual users within that domain then generate their own public/private key pairs and certificate signed by themselves. These certificates are then stored securely by a trusted server within that domain. When the certificate is retrieved, the retriever can validate that the certificate was obtained securely from a valid server within the domain, and hence have some assurance that this is a valid certificate for a particular user within the domain. The IETF SACRED (Securely Available Credentials) working group [26] has developed methods for accessing and transport credentials such as private keys. This is described in some detail in Chapter 11.

2.3.2 Certificate Authorities

A certificate authority issues a certificate and signs it with its private key. If a recipient of the certificate trusts the CA, it can validate the certificate using the CA's public key. Figures 2.9 and 2.10 show the chain of trust used in the certificate in Figure 2.8. Note that the www.artechhouse.com certificate was issued by

Serial Number:
78:EE:48:DE:18:5B:20:71:C9:C9:C3:B5:1D:7B:DD:C1

Certificate Signature Algorithm:
PKCS #1 SHA-1 With RSA Encryption

Issuer:
OU = Class 3 Public Primary Certification Authority
O = VeriSign, Inc.
C = US

4/16/1997 19:00:00 PM
(4/17/1997 0:00:00 AM GMT)

10/24/2011 18:59:59 PM
(10/24/2011 23:59:59 PM GMT)

Subject:
OU = www.verisign.com/CPS Incorp.by Ref. LIABILITY LTD.(c)97 VeriSign
OU = VeriSign International Server CA - Class 3
OU = VeriSign, Inc.
O = VeriSign Trust Network

Subject Public Key Algorithm:
PKCS #1 RSA Encryption

Subject Public Key:
30 81 89 02 81 81 00 d8 82 80 e8 d6 19 02 7d 1f
85 18 39 25 a2 65 2b e1 bf d4 05 d3 bc e6 36 3b
aa f0 4c 6c 5b b6 e7 aa 3c 73 45 55 b2 f1 bd ea
97 42 ed 9a 34 0a 15 d4 a9 5c f5 40 25 dd d9 07
c1 32 b2 75 6c c4 ca bb a3 fe 56 27 71 43 aa 63
f5 30 3e 93 28 e5 fa f1 09 3b f3 b7 4d 4e 39 f7
5c 49 5a b8 c1 1d d3 b2 8a fe 70 30 95 42 cb fe
2b 51 8b 5a 3c 3a f9 22 4f 90 b2 02 a7 53 9c 4f
34 e7 ab 04 b2 7b 6f 02 03 01 00 01

Certificate Signature Algorithm:
PKCS #1 SHA-1 With RSA Encryption

Certificate Signature Value:
23 5d ee a6 24 05 fd 76 d3 6a 1a d6 ba 46 06 aa
6a 0f 03 90 66 b2 b0 a6 c2 9e c9 1e a3 55 53 af
3e 45 fd dc 8c 27 dd 53 38 09 bb 7c 4b 2b ba 95
4a fe 70 4e 1b 69 d6 3c f7 4f 07 c5 f2 17 5a 4c
a2 8f ac 0b 8a 06 db b9 d4 6b c5 1d 58 da 17 52
e3 21 f1 d2 d7 5a d5 e5 ab 59 7b 21 7a 86 6a d4
fe 17 11 3a 53 0d 9c 60 a0 4a d9 5e e4 1d 0c 29
aa 13 07 65 86 1f bf b4 c9 82 53 9c 2c 02 8f 23

Figure 2.9 Next-level certificate.

```
Serial Number:
70:BA:E4:1D:10:D9:29:34:B6:38:CA:7B:03:CC:BA:BF

Certificate Signature Algorithm:
PKCS #1 MD2 With RSA Encryption

Issuer:
OU = Class 3 Public Primary Certification Authority
O = VeriSign, Inc.
C = US

1/28/1996 18:00:00 PM
(1/29/1996 0:00:00 AM GMT)

8/1/2028 18:59:59 PM
(8/1/2028 23:59:59 PM GMT)

Subject:
OU = Class 3 Public Primary Certification Authority
O = VeriSign, Inc.
C = US

Subject Public Key Algorithm:
PKCS #1 RSA Encryption

Subject's Public Key:
30 81 89 02 81 81 00 c9 5c 59 9e f2 1b 8a 01 14
b4 10 df 04 40 db e3 57 af 6a 45 40 8f 84 0c 0b
d1 33 d9 d9 11 cf ee 02 58 1f 25 f7 2a a8 44 05
aa ec 03 1f 78 7f 9e 93 b9 9a 00 aa 23 7d d6 ac
85 a2 63 45 c7 72 27 cc f4 4c c6 75 71 d2 39 ef
4f 42 f0 75 df 0a 90 c6 8e 20 6f 98 0f f8 ac 23
5f 70 29 36 a4 c9 86 e7 b1 9a 20 cb 53 a5 85 e7
3d be 7d 9a fe 24 45 33 dc 76 15 ed 0f a2 71 64
4c 65 2e 81 68 45 a7 02 03 01 00 01

Certificate Signature Algorithm:
PKCS #1 MD2 With RSA Encryption

Certificate Signature Value:
bb 4c 12 2b cf 2c 26 00 4f 14 13 dd a6 fb fc 0a
11 84 8c f3 28 1c 67 92 2f 7c b6 c5 fa df f0 e8
95 bc 1d 8f 6c 2c a8 51 cc 73 d8 a4 c0 53 f0 4e
d6 26 c0 76 01 57 81 92 5e 21 f1 d1 b1 ff e7 d0
21 58 cd 69 17 e3 44 1c 9c 19 44 39 89 5c dc 9c
00 0f 56 8d 02 99 ed a2 90 45 4c e4 bb 10 a4 3d
f0 32 03 0e f1 ce f8 e8 c9 51 8c e6 62 9f e6 9f
c0 7d b7 72 9c c9 36 3a 6b 9f 4e a8 ff 64 0d 64
```

Figure 2.10   Top-level certificate—self-signed certificate.

the "VeriSign International Server CA - Class 3" whose certificate was issued by "Class 3 Public Primary Certification Authority." The top level of the chain is certificate of the "Class 3 Public Primary Certification Authority" whose certificate was issued by "Class 3 Public Primary Certification Authority." This last certificate is an example of a self-signed certificate. The top level certificate in any trust hierarchy will always be a self-signed certificate, by definition.

A recipient can trust more than one CA, and it can organize that trust in a hierarchy, as shown in Figure 2.11.

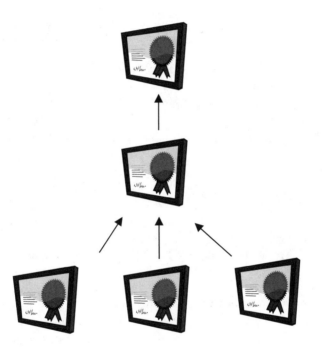

Figure 2.11   Certificate authority trust hierarchy.

References

[1]   Schneier, B., *Applied Cryptography*, 2nd Ed., New York: John Wiley & Sons, 1996.

[2]   Smith, R., *Internet Cryptography*, Addison Wesley, 1997.

[3]   ANSI X3.106, "American National Standard for Information Systems-Data Link Encryption," American National Standards Institute, 1983.

[4]   ANSI X9.52-1998, "Triple Data Encryption Algorithm Modes of Operation," American National Standards Institute, 1998.

[5] Lai, X., *On the Design and Security of Block Ciphers,* ETH Series in Information Processing, V. 1, Konstanz: Hratung-Gorre Verlag, 1992.

[6] Advanced Encryption Standard (AES), Federal Information Processing Standard 197, November 2001, http://csrc.nist.gov/publications/fips/fips197/fips-197.pdf.

[7] Rivest, R.L., *The RC4 Encryption Algorithm,* RSA Data Security, Inc., April 1992.

[8] Kaufman, C., R. Perlman, and M. Speciner, *Network Security: Private Communications in a Public World,* Upper Saddle River, NJ: Prentice Hall, 1995.

[9] Rivest, R.L., A. Shamir, and L.M. Adleman, "The MD5 Message Digest Algorithm," RFC 1321, April 1992.

[10] NIST, FIPS PUB 180, "Secure Hash Standard," U.S. Department of Commerce, May 1993.

[11] Rivest, R., A. Shamir, and L. M. Adleman, "A Method for Obtaining Digital Signatures and Public-Key Cryptosystems," *Communications of the ACM,* Vol. 21, No. 2, February 1978, pp. 120–126.

[12] U.S. National Institute of Standards (NIST), FIPS PUB 186: Digital Signature Standard (DSS), May 1994.

[13] Krawczyk, H., M. Bellare, and R. Canetti, "HMAC: Keyed-Hashing for Message Authentication," RFC 2104, February 1997.

[14] Rivest, R.L., A. Shamir, and L.M. Adleman, "A Method for Obtaining Digital Signatures and Public-Key Cryptosystems," *Communications of the ACM,* Vol. 21, No. 2, February 1978, pp. 120–126.

[15] Diffie, W., and M.E. Hellman, "New Directions in Cryptography," *IEEE Transactions on Information Theory,* Vol. 22, 1976, pp. 644–84.

[16] Diffie, W., and M.E. Hellman, "Authenticated Key Exchange and Secure Interactive Communication," *Proceedings of SECURICOM '90,* 1990.

[17] Chokani, S., and W. Ford, "Internet X.509 Public Key Infrastructure Certificate Policy and Certification Practices Framework," RFC 2527, March, 1999.

[18] Piscitello, D., Understanding Certificates and PKI, Core Competence, http://www.corecom.com/external/livesecurity/pki.htm.

[19] Housley, R., and T. Polk, *Planning for PKI,* New York: John Wiley & Sons, 2001.

[20] Choudhury, S., K. Bhatnagar, and W. Hague, *Public Key Infrastructure: Implementation and Design,* New York: John Wiley & Sons, 2002.

[21] Oppliger, R., *Contemporary Cryptography,* Norwood, MA: Artech House Publishers, 2005.

[22] Housley, R, "Internet X.509 Public Key Infrastructure and CRL Profile," RFC 2459, November 1998.

[23] Myers, M., R. Ankney, A. Malpani, S. Galperin, and C. Adams, "X.509 Internet Public Key Infrastructure Online Certificate Status Protocol—OCSP," RFC 2560, June 1999.

[24] PKIX Public Key Infrastructure IETF Working Group, http://www.ietf.org/pkix.html.

[25] Freeman, T., R. Housley, A. Malpani, D. Cooper, and T. Polk, "Simple Certificate Validation Protocol," IETF Internet-Draft, work in progress, August 2005.

[26] SACRED (Securely Available Credentials) IETF Working Group, http://www.ietf.org/sacred.html.

3

VoIP Systems

3.1 Introduction

In this chapter, we present the components and protocols of VoIP systems. We reference the terminology and basics presented in this chapter throughout the book. For a more complete and detailed examination of VoIP systems in general, and the Session Initiation Protocol (SIP) specifically, refer to [1] and [2].

3.1.2 VoIP Architectures

Voice over Internet Protocol is one of many telephony applications that can be used over Internet Protocol (IP). By operating VoIP protocols over Internet Protocol, IP users can place voice, fax, and modem calls over IP-based networks. VoIP uses a signaling protocol to establish sessions (conversations). Once VoIP endpoints agree on the parameters for a call, all media signals between the endpoints are digitized, compressed, "packetized," and exchanged as IP datagrams.

VoIP is a very different application from traditional Internet applications, but ultimately, it is simply another application. The relation between VoIP and the Internet (TCP/IP) architecture is illustrated in Figure 3.1.

In Figure 3.1, we illustrate the Session Initiation Protocol as the VoIP signaling protocol. Other signaling protocols, both proprietary and internationally standardized, exist. SIP, however, is emerging as the dominant signaling protocol and is the main focus of this book.

The principle VoIP protocols include:

- Session Initiation Protocol (SIP, RFC 3261), a call signaling protocol [3].

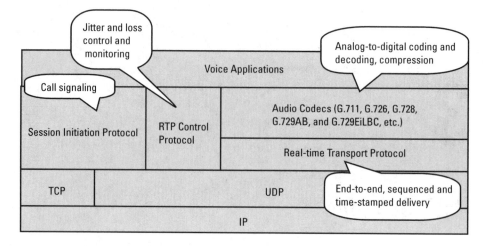

Figure 3.1 Relationship between VoIP and TCP/IP architecture.

- Real-time Transport Protocol (RTP, RFC 3550), an end-to-end delivery protocol for real-time media applications.

- RTP Control Protocol (RTPC, RFC 3550), part of the RTP specification that enables call quality monitoring (jitter, packet loss, latency).

- Session Description Protocol (SDP, RFC 2327), a protocol that provides the means to specify the characteristics of a media session [142].

- Codecs for the encoding and decoding of analog signals into digital signals (compressors/decompressors, or codecs).

- ITU-T H series recommendations for audiovisual and multimedia systems.

In Figure 3.1, you may note that VoIP media sessions run over RTP rather than directly over one of the Internet's existing transport protocols, TCP and UDP. After considerable early experimentation with voice over IP, the research community concluded that TCP/IP wasn't entirely suitable for real-time applications. TCP's retransmission strategies can introduce high delays and cause delay jitter. In addition, its commonly implemented congestion control mechanism (Jacobsen slow start and congestion avoidance) was poorly suited for media sessions, and TCP did not support multicast service, which is useful for conferencing. UDP alone was deemed unsuitable because there is no defined technique for synchronizing UDP stream, so streams from different servers could collide. Even if present, a feedback channel, defined for quality control, was required. RTP (and RTCP) were defined to provide service elements needed for VoIP and to operate above Internet's transport layer.

3.2 Components

The basic components of a VoIP system are shown in Figure 3.2. They include:

- VoIP endpoints (also called user agents)—the applications and devices making and receiving VoIP calls. These devices can be software clients running on laptops, PCs, or PDAs, or they can be standalone devices such as IP telephones or adapters that allow a PSTN telephone to be plugged in and used.

- Servers—provide various functionality in a VoIP system. The type and nature of the servers is protocol-dependent.

- Infrastructure servers, including domain name system (DNS) and authentication, authorization, and accounting (AAA) servers—provide these key functions in a network.

- VoIP application servers—provide enhanced "telephony" services such as voicemail, conferencing, messaging, and interactive voice response (IVR) systems.

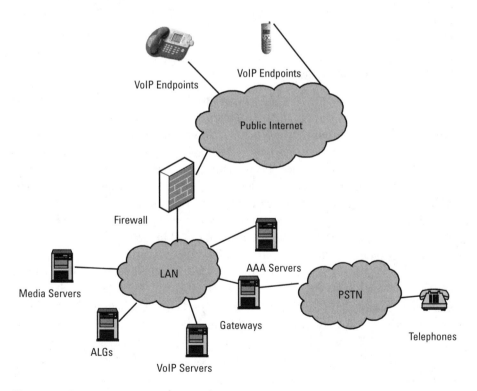

Figure 3.2 Components of a VoIP system.

- Media servers—play prompts, announcements, and perform mixing. For example, a conference bridge takes in a multiple media streams and combines them into one or more output streams.

- Proxy servers—facilitate call control (signaling) between user agents and call routing. Proxy servers can also provide registration and redirection services as defined in SIP.

- Gateways—provide interworking between VoIP networks and protocols and other voice networks and protocols. Gateways to the PSTN are a critical component, and aspects of their security are discussed in detail in Chapter 12.

- Security devices—include firewalls, application layer gateways (ALGs) and other systems which provide security functions.

3.3 Protocols

This section offers a brief overview of various protocols used in VoIP systems that will be analyzed throughout the rest of the book. For a more detailed explanation of the operation of these protocols, a number of references are provided at the end of this chapter.

3.3.1 Session Initiation Protocol

Session Initiation Protocol (SIP) [3] is the dominant standard signaling protocol for VoIP systems. Developed by the Internet Engineering Task Force (IETF), the standards body for Internet protocols, it has an Internet heritage and draws on many aspects of Internet architecture and applications, not the least of which are generalized Internet security mechanisms. SIP was first published as RFC 2543 [5] in 1999, then updated with RFC 3261 in 2003. The later specification is backwards compatible with the earlier specification and utilizes the same version number, version 2.0. Many of the improvements and upgrades in RFC 3261 relate to security and will be discussed in detail in later chapters.

An example SIP call flow is shown in Figure 3.3.

The basic SIP request or method, INVITE, is used to invite another participant into a media session. Responses in SIP are numerical, with the first digit indicating the class. The 100 Trying is a single hop response which is never forwarded by servers. It is simply used to suppress retransmissions of SIP messages when UDP transport is used. The 180 Ringing response indicates that the called party is being alerted and is often used to generate ringback tone to the caller.

All responses, with the exception of 100 Trying, are routed back through the same set of proxy servers as the request. This is done through the

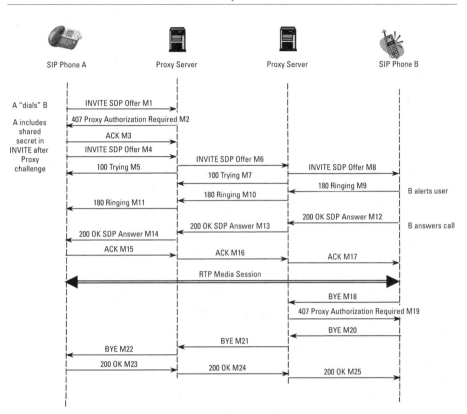

Figure 3.3 Basic SIP call flow.

Via header field chain which puts an IP address and port number at which it is listening for a response.

An example SIP message, from Figure 3.3, follows.

```
INVITE sip:bob@voiptheworld.net SIP/2.0
To: <sip:bob@voiptheworld.net>
From: "Alice" <sip:alice@ipislands.com>;tag=f1147271
Via: SIP/2.0/UDP206.65.230.170:64032;branch=z4bK-d8;rport
Call-ID: b70022781871ee3e@bWNpLWU2ejIwNTl2YWdiLm1jaW
CSeq: 1 INVITE Contact: <sip:alice@206.65.230.170:64032>
Max-Forwards: 70
Allow: INVITE, ACK, CANCEL, OPTIONS, BYE, REFER, NOTIFY,
  MESSAGE, SUBSCRIBE, INFO
Content-Type: application/sdp
Content-Length: 403

v=0
o=- 1313802769 1313803240 IN IP4 206.65.230.170
s=-
```

```
c=IN IP4 206.65.230.170
t=0 0
m=audio 64028 RTP/AVP 98 100 0 8 18 101
a=fmtp:101 0-15
a=rtpmap:100 speex/16000
a=rtpmap:98 ilbc/8000
a=rtpmap:101 telephone-event/8000
a=sendrecv
```

The SIP request begins with the method name, INVITE, a space, the destination address (e.g., a uniform resource identifier, URI) for this request, known as the Request-URI, a space, and the current version of SIP, SIP/2.0. In this example, the Request-URI is sip:bob@voiptheworld.net. After this first line, almost anywhere in a SIP message that a single space is permitted, multiple spaces or tabs are also permitted. SIP header fields can be folded onto multiple lines if the continuation line begins with a space or a tab.

SIP implementations use the UDP port 5060, which is assigned by IANA from the well known port space (http://www.iana.org/assignments/port-numbers). Servers typically listen on this port number for incoming requests. Port 5060 is the default port. Other ports can be used by explicitly including the port number in the SIP URI. For example, to use port 13453, one would encode the SIP URI as follows:

```
sip:bob@pc43.voiptheworld.net:13453
```

SIP clients can utilize a different port than 5060 by registering a port number other than 5060 in the Contact URI, for example:

```
sip:alice@206.65.230.170:64032.
```

Secure SIP, described in [3] uses port 5061.

SIP uses two types of URIs: an Address of Record (AOR) or a device URI. An Address of Record URI is usually associated with a user or a service. It is generally a long-lived identifier, suitable for publishing in a directory, printing on a business card, or posting on a web page. A Contact or device URI is associated with a particular device or endpoint. A device URI can be temporarily associated with a particular user using the SIP REGISTER method, as shown in Figure 3.4.

Following is an example of a SIP REGISTER request:

```
REGISTER sips:registrar.biloxi.com SIP/2.0
Via: SIP/2.0/TLSbobspc.biloxi.com:13423;branch=z9hG4bKna
Max-Forwards: 70
To: Bob <sips:bob@biloxi.com>
```

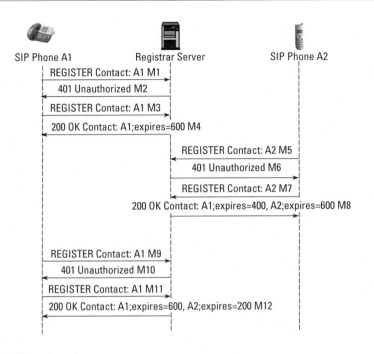

Figure 3.4 SIP registration.

```
From: Bob <sips:bob@biloxi.com>;tag=456248
Call-ID: 843817637684230@998sdasdh09
CSeq: 1826 REGISTER
Contact: <sips:bob@192.0.2.4:13423>
Expires: 600
Content-Length: 0
```

In this request, the AOR URI sips:bob@biloxi.com is temporarily bound (e.g., for 10 minutes or 600 seconds) to the Contact URI sips:bob@192.0.2.4:13423. In the example, Bob's device is listening on port 13423 instead of the default port 5060, as indicated in the Via header field and also the Contact URI.

The 200 OK response is sent to indicate an answer condition. An ACK is sent in response to a final (non-1xx) response. The ACK is considered a separate transaction, but it always is sent in response to an INVITE transaction and never any other request type transaction. That is, an OPTIONS, REFER, or BYE transaction never generates an ACK transaction. Also note that no responses are sent to ACKs.

SIP has its own built in reliability mechanisms and does not require TCP. Retransmission timers allow SIP to operate over a best effort transport such as UDP. In general, a request is retransmitted until a final response is received. The retransmission timers in SIP use an exponential back-off mechanism.

The Command Sequence (CSeq) header field is used to distinguish a retransmission from a new request. Each new request increments the CSeq number, while a retransmission keeps the same CSeq number. The exceptions to this are ACK and CANCEL, which use the same CSeq number as the INVITE transaction that they reference. That is, for an INVITE with CSeq: 3424 INVITE, an ACK for this transaction would have CSeq: 3424 ACK, or a CANCEL would have a CSeq: 3424 CANCEL.

The set of SIP response classes follows:

- A 1xx response is a provisional or informational response, providing information about the progress of the request but does not say whether it ultimately has succeeded or failed.

- A 2xx response is a success response, indicating that the request has completed or succeeded.

- A 3xx response is a redirection response, redirecting the request to another destination contained in the request.

- A 4xx response is a client error response, indicating that the request has failed. Information in the response or the exact response code indicates why the request failed and how the request can be fixed or retried.

- A 5xx response is a server error response, indicating that the request has failed due to a problem with the server. The response may contain information about when or where the request can be retried.

- A 6xx response is a global failure response, indicating that the request has failed at this location and will fail at all other locations.

SIP borrows an ASCII-based structure from Hyper Text Transport Protocol (HTTP) [6]. SIP messages are constructed using a set of header fields (similar to those in an e-mail message such as To, From, Subject, etc.) then possibly one or more message bodies.

SIP is an application protocol, and can operate over a variety of underlying transport and secure delivery protocols including UDP (User Datagram Protocol) [7], TCP (Transmission Control Protocol) [8], SCTP (Stream Control Transport Protocol) [9], and TLS (Transport Layer Security) [10]. SIP will likely be extended to utilize DTLS transport [11].

Endpoints in SIP are known as user agents (UAs). UAs can be applications or clients running on PCs, laptops, or PDAs. They can be firmware running in an embedded device such as a SIP phone, telephone adapter, or wireless LAN (WiFi) SIP phone. SIP UA software can also be included in applications and used in call control, signaling, directory lookup, and other applications.

SIP, used in many commercial VoIP offerings, is available for Windows, Macintosh, and open source operating systems, and is part of many enterprise instant messaging systems. SIP is also the call signaling, presence, and instant messaging protocol in 3GPP (Third Generation Partnership Project) cellular telephone systems [12]. In addition to its use in session establishment in VoIP, SIP can be used for video, gaming, conferencing and collaboration, and also presence and instant messaging (IM) systems. In fact, the first standards based presence and IM gateways between the largest four consumer providers have been implemented using SIP [13].

SIP systems utilize a number of types of servers. These servers are listed in Table 3.1. The most common server is a proxy server, and this server will be discussed throughout the book.

Voice applications commonly use SIP to establish a media session, either audio or voice. Other media types include video and text, and SIP is easily extended to support future media types. The media session information is carried in an offer/answer mechanism for simple capability negotiation, in a Session Description Protocol (SDP) [14] message body. The offer is usually in the INVITE and the answer in the 200 OK response. Other combinations are possible. For example, the offering party can let the answering party decide what type of session to establish. In this case, the offer is in the 200 OK and the answer is in the ACK. Using other SIP methods, it is possible to have offers and answers in reliable 183 responses, as well as other SIP methods.

Additional SIP Call flows are shown in Figures 3.4, 3.5, 3.6, 3.7, and 3.8. Figure 3.5 shows a user agent registering using a REGISTER request, then querying the capabilities of another UA using an OPTIONS request.

Table 3.1
Types of SIP Servers

Server Type	Function
Registrar	Accepts registration requests for dynamic IP address mapping
Proxy	Receives and forwards requests—SIP layer router
Redirect	Receives requests and returns redirection response (3xx)
Location Service	Used by Registrar to store registration data, used by Proxy to retrieve SIP routing information for incoming requests
Identity Service	Provides a cryptographic identity assurance (described in detail in Chapter 11)
Certificate Service	Provides storage and retrieval of self-signed certificates and credentials for establishing secure sessions (described in detail in Chapter 11)

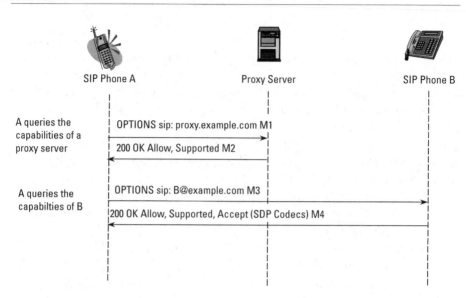

Figure 3.5 Querying of capabilities.

A response to an OPTIONS request can contain the following information:

- SIP request types (methods);

- SIP event packages supported;

- Message body MIME types supported;

- Features and SIP extensions supported;

- Codecs and media types supported.

CANCEL is a special case in SIP and is done on a hop-by-hop basis, which makes authentication of CANCEL very difficult. Figure 3.6 shows the use of a CANCEL request by a proxy server to implement a sequential search for a single user among several devices.

This example assumes that both of these devices have registered against the same user's AOR.

Figure 3.7 shows how the REFER method can be uses to transfer a call from one party to another. In this example, we show a simple, unattended transfer operation.

More complex transfers can be done using SIP as detailed in [14]. For simplicity, Figure 3.6 does not illustrate proxy servers or authentication challenges.

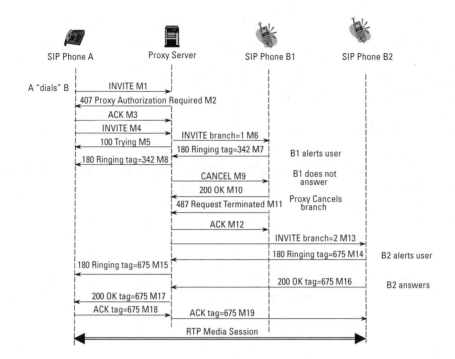

Figure 3.6 Canceled session.

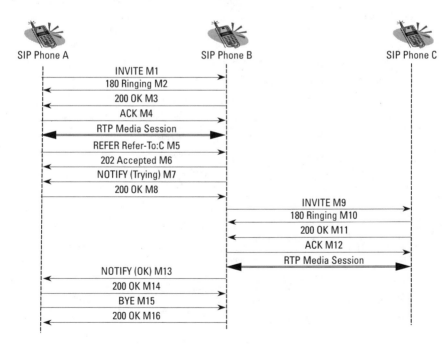

Figure 3.7 Transferred session.

Figure 3.8 shows a presence and instant messaging call flow for SIP. This example illustrates how SIP can be utilized for much more than just VoIP.

Many of the security aspects discussed in this book apply to presence and instant messaging services. However, these services introduce additional security issues that are not discussed in this book.

3.3.2   Session Description Protocol

Session Description Protocol (SDP, RFC 2327) is used to describe media sessions. Typical information provided in an SDP message body includes the type of session, media codec, IP address and port number for receiving data. An example SDP message follows:

```
v=0
o=- 1313802769 1313803240 IN IP4 206.65.230.170
s=-
c=IN IP4 206.65.230.170
=0 0
m=audio 64028 RTP/AVP 100 0 8 18 98 101
```

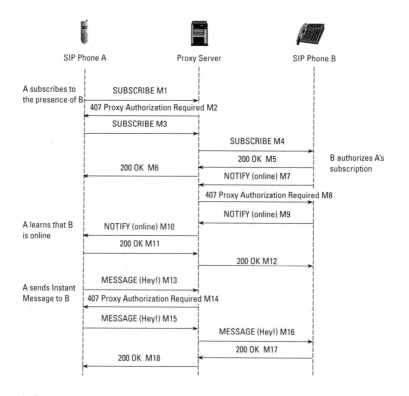

Figure 3.8 Presence and instant messaging session.

```
a=rtpmap:100 speex/16000
a=rtpmap:98 ilbc/8000
a=rtpmap:101 telephone-event/8000
a=sendrecv
```

SDP fields are in required order, as listed in Table 3.2.

The media line (m=) carries most of the information of interest to a VoIP system. It indicates the media type (audio, video, text, application, and message are defined), the port number that the application is listening on, the profile type (RTP/AVP, RTP/SAVP, or RTP/SAVPF), and a list of payload types.

The use of SDP with SIP, known as the Offer/Answer protocol, is defined in RFC 3264 [15].

The use of the encryption key field (k=) in SDP has been deprecated. Other SDP extensions to carry keying material are described in Chapter 10.

SDP has no built-in security, but instead, relies on the security of the application which uses or transports it. The use of TLS transport with SIP can provide hop-by-hop confidentiality and integrity protection for SDP. However, TLS can

Table 3.2
SDP Fields in Required Order

Field	Name	Mandatory/ Optional
v =	version number	m
o =	owner/creator of session identifier	m
s =	session name	m
i =	session information	o
u =	uniform resource identifier	o
e =	email address	o
p =	phone number	o
c =	connection information	m
b =	bandwidth information	o
t =	session start and stop time	m
r =	repeat times	o
z =	time zone corrections	o
a =	attribute lines	o
m =	media lines	o
c =	media connection information	o
a =	media attributes	o

only provide end-to-end protection if there are no Proxy Servers between the communicating UAs. If there is a proxy server in the path, two separate TLS connections are needed: one from the initiating UA to the proxy server, and one from the proxy server to the called UA (or voice application server, e.g., a voice mail system). As a result of this scenario, the SDP information will be available (exposed) to the proxy server. S/MIME can provide end-to-end confidentiality and integrity protection, as discussed in Chapter 9.

3.3.3   H.323

H.323 [16], or Packet-Based Multimedia Communication, is a protocol developed by the International Telecommunication Union (ITU-T). The first version of H.323 was adopted in 1996, and the most recent version (5) was adopted in 2003. Version 4 is currently the most widely deployed version, used primarily in ISDN-based video conferencing systems and international toll bypass VoIP applications. All the H.323 versions are backwards compatible with earlier versions, and the version used is negotiated during the session establishment.

The basic elements of an H.323 network include:

- Endpoints—also called terminals;
- Gatekeepers, which provide signaling and services;
- Multipoint control units (MCUs), which provide conferencing services;
- Gateways, which provide interworking with other protocols such as PSTN and SIP.

H.323 Gatekeeper discovery and terminal registration is shown in Figure 3.9. The Registration Admission and Status (RAS) protocol, H.225, provides Gatekeeper discovery. A Gatekeeper Request (GRQ) is sent using multicasting over the local network. If the local Gatekeeper responds with a Gatekeeper Confirmation (GCF) response, the terminal can register with the Gatekeeper using the Registration Request (RRQ) and Registration Confirmation (RCF) messages.

A basic call setup using FastStart shown in Figure 3.10 below. Before placing a call, an H.323 terminal sends an Admission Request (ARQ) and receives an Admission Confirmation (ACF) from a Gatekeeper using RAS. A Setup request is sent which contains the capabilities of the calling terminal to the destination terminal.

H.323 references a number of other protocols, as indicated in Table 3.2. H.225 is used for registration, admission, and call signaling, as shown in Figures 3.9 and 3.10. H.245 is used to establish and control the media sessions, and

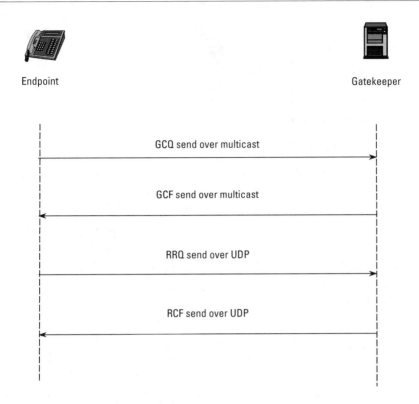

Endpoint   Gatekeeper

GCQ send over multicast

GCF send over multicast

RRQ send over UDP

RCF send over UDP

Figure 3.9 H.323 registration example.

H.235 describes how keys can be exchanged and media encrypted for security. T.120 is used for conferencing applications in which a shared whiteboard application is used. The G.7xx series of specifications defines audio codecs used by H.323, while the H.26x series of specifications defines the video codecs. Like SIP, H.323 uses RTP for media transport and RTCP for control of the RTP sessions. H.450 defines supplementary telephony services for H.323 such as call transfer, call hold and so forth.

H.323 uses multicast for Gatekeeper discover, UDP port 1719 for RAS, TCP port 1720 for the call control channel (Q.931). If used, H2.45 call control channel dynamically assigns an ephemeral port number.

H.323 and SIP can interwork, using an available number of commercial gateways. The basic requirements are listed in [17]. The use of H.323 URIs and the use of H.323 with ENUM will be described in later chapters. A H.323 URI can be used by an endpoint to locate an appropriate Gatekeeper for routing an H.323 request to the destination.

H.323 Security is defined in the H.235 specification. The current version of H.235 is version 3, approved in 2003. H.235 specifies how H.323 terminals and Gatekeepers can use TLS or IPSec for transport security. H.235 also defines

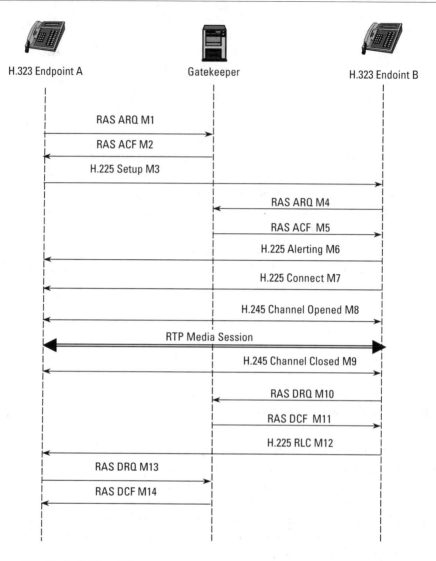

Figure 3.10 Basic H.323 call flow.

mechanisms for media encryption key exchange over H.245, and the generation of an encryption key using a SHA-1 hash of a user's passphrase.

Although there are many H.323 VoIP systems deployed today, virtually all new standard VoIP deployments are based on SIP. Existing H.323 VoIP systems tend to have extremely low levels of security, with no encryption or authentication beyond source IP address filtering. Most of the current work on VoIP security and advanced services and systems is being built around SIP, and therefore, discussion of SIP security will be our main focus.

Table 3.2

H.323 Referenced Protocols

Protocol	Description
H.225	Registration, Admission, and Status (RAS) and call signaling
H.245	Control signaling (media control)
H.235	Security
T.120	Multipoint graphic communication (white board)
G.7xx	Audio codecs
H.26x	Vide codecs
RTP	Real Time Transport Protocol, RFC 3550
RTCP	RTP Control Protocol, RFC 3550
H.450	Supplementary Services

3.3.4 Media Gateway Control Protocols

A media gateway provides the media connectivity between the PSTN and the IP networks by converting the audio channel into RTP packets for transport over IP. The media gateway controller is a signaling element that is interposed between and interacts with the telephony signaling protocols of the PSTN and IP signaling protocols such as SIP or H.323. A number of master/slave protocols are used between an MG and an MGC. Many commercial VoIP gateways can be decomposed using one or more of these protocols. A number of media gateways can be controlled from a centralized MGC as shown in Figure 3.11. Note that a MGC which also provides a variety of services is sometimes known as a softswitch.

Two standard media gateway control protocols and several and proprietary protocols are currently in use. Media Gateway Control Protocol (MGCP) [18] was first published as an informational rather than standard protocol specification by the IETF in 1998. MGCP was updated in 2003 but remains as an informational RFC.

MEGACO [19] was developed jointly between the IETF and the ITU-T and published as both RFC 3015 and H.248. MEGACO is considerably more complex than MGCP and also uses extensible packages, a number of which have been standardized.

All gateway decomposition protocols are similar in that they are typically part of a service provider's infrastructure, and run within a single administrative domain. Securing them is much simpler than actual interdomain signaling protocols such as SIP and H.323. Typically, encryption integrity protection and

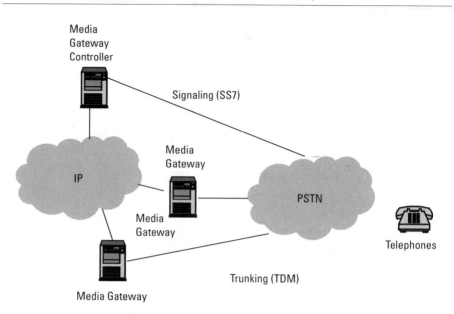

Figure 3.11 Decomposed media gateway/media gateway controller architecture.

authentication provided by a security protocol provides sufficient security (see Chapter 6).

3.3.5   Real Time Transport Protocol

The Real Time Transport Protocol (RTP) [20] was first published in 1996 as RFC 1889, with profiles for audio and video conferences defined in RFC 1990. RTP profiles for audio and video conferences with minimal control [21] define the basic characteristics of a media session such as the codec and sampling rate. The current version of RTP and the companion protocol RTP Control Protocol (RTCP) [20] are the mainstays of VoIP real-time media communications, used by virtually every VoIP system, including VoIP systems that utilize proprietary signaling protocols.

RTCP is a secondary communication channel, used to exchange information about the RTP session including quality statistics. During silence suppression or when placed on hold, there may be no RTP traffic but RTCP traffic will continue between the endpoints.

RTP uses ephemeral UDP ports negotiated by the call signaling protocol. For example, with SIP, RTP ports are carried in SDP message bodies. For H.323, the UDP ports are carried in the H.245 call control channel.

Additional details of RTP and its secure profile SRTP (Secure RTP) will be discussed in detail in Chapter 10.

3.3.6 Proprietary Protocols

A number of proprietary VoIP signaling protocols are used today. Some are master/slave in nature and have a similar architecture. Their closed nature does not allow a proper security analysis of their architectures. However, observation and analysis of proprietary traffic flows provides some insight into their operation, such as how network address translation and firewall traversal are implemented. These observations suggest that many proprietary signaling protocols incorporate very little security and do not employ encryption methods described in Chapter 2.

Some proprietary signaling applications such as Skype [22] use a peer-to-peer architecture. Skype users join as members of a peer-to-peer (P2P) network to establish VoIP sessions as well as presence and IM services. The Skype architecture utilizes *supernodes* to provide login, authentication, directory searches, and NAT and firewall traversal, as shown in Figure 3.12.

As Skype uses an unpublished proprietary protocol for signaling and media, we cannot provide a complete security analysis. However, some security information about the service is becoming available [24, 25]. This information indicates that the signaling is encrypted using RC4 and the media using AES. Authentication is done using certificates issued by a centralized CA. Key agreement for the media sessions is done using a proprietary scheme in which

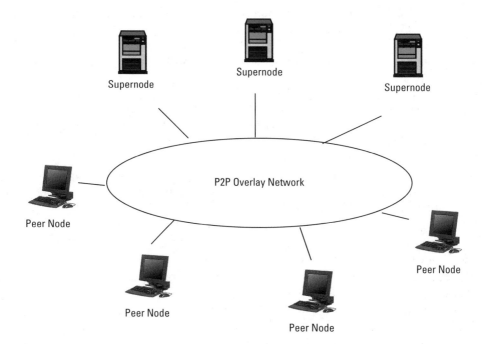

Figure 3.12 Peer-to-peer architecture with supernodes.

each participant contributes 128 bits towards a 256 bit key. For a proper and thorough review of its security, Skype should publish the details of their algorithms.

Note that SIP already has many P2P aspects. Current work is underway to investigate how some of the functionality provided by SIP servers can be moved to peer nodes for a complete P2P architecture [25–27].

3.4   Security Analysis of SIP

This section will perform a basic security analysis of SIP. More detailed analyses and consideration of additional issues is discussed in later chapters.

SIP Requests can be authenticated using a number of mechanisms:

A SIP INVITE can be challenged to provide a shared secret using the HTTP Digest mechanism (Chapter 9). A request resent after a challenge will have an incremented CSeq number, since SIP uses the CSeq number to distinguish between a retransmission and a new request. Since ACK and CANCEL cannot increment the CSeq number, they can not be challenged using HTTP Digest.

A SIP INVITE request can be authenticated if it comes in over a connection that has previously been authenticated (e.g., when a SIP INVITE request is transmitted over a TLS connection where client subauthentication or mutual authentication has been performed).

"Loose authentication" of a SIP INVITE can be performed over UDP and TCP by verifying that the source IP and port number match ones previously authenticated. Because this method is vulnerable to address spoofing, it should be avoided.

The SIP INVITE request may have a cryptographic signature that can be verified. This is discussed in Chapter 11 in terms of the Identity header field and Secure Multipurpose Internet Mail Exchange or S/MIME.

SIP responses cannot be challenged using the first listed method. Mutual authentication is possible using the rarely implemented Authentication-Info header field. This form of authentication requires a second shared secret between the server and the client.

The inability to challenge ACK and CANCEL is seen as a weakness in the protocol specification. An ACK can contain an SDP answer, and the injection of a false ACK can cause an incorrect media session to be established. An injected CANCEL could be used in a DoS attack to terminate pending calls. The attacker must be able to inspect the SIP requests to construct a false request that matches the correct dialog (To, From, and Call-ID spoofed correctly) to successfully execute both attacks. The use of confidentiality services in a supporting protocol such as TLS or IPSec makes these attacks more difficult.

If proxy servers do not Record-Route and drop out of the session as shown in Figure 3.2, any chain of trust established and used for the INVITE transaction is not available for the ACK and subsequent requests.

If integrity protection is not available a REGISTER could be modified between a UA and the Registrar server by changing the Contact URI. This would result in incoming sessions being hijacked by the attacker. An INVITE could be modified so that the media stream is diverted to a recording device, or to another person.

Most of these attacks are prevented by utilizing a transport with integrity protection as discussed in Chapter 9.

One approach for dealing with unauthenticated CANCEL requests is to ignore them. This will result in sessions being established, but the BYE will be automatically sent and can be properly authenticated. Unless the ACK is carrying an SDP answer, injecting false ones has limited value for an attacker. Since some implementations will automatically send a BYE if a malformed ACK is received, injecting malformed ACKs in a dialog could result in session disruption.

The confidentiality of certain signaling information is also important otherwise certain SIP call control operations can be forged by an attacker. For example, the combination of the To tag, From tag, and Call-ID is known as the dialog indentifier. The Replaces header field [28] can be used to take over an existing session, but only if the dialog identifiers are known.

References

[1]   Johnston, A. B., *SIP: Understanding the Session Initiation Protocol*, 2nd Ed., Norwood, MA: Artech House, 2004.

[2]   Sinnreich, H., A.B. Johnston, and R.J. Sparks, *SIP Beyond VoIP*, New York: VON Publishing, 2005.

[3]   Rosenberg, J., H. Schulzrinne, G. Camarillo, A. Johnston, J. Peterson, R. Sparks, M. Handley, and E. Schooler, "SIP: Session Initiation Protocol," RFC 3261, June 2002.

[4]   Handley, M., and V. Jacobson, "SDP: Session Description Protocol," RFC 2327, April 1998.

[5]   Handley, M., H. Schulzrinne, E. Schooler, and J. Rosenberg, "SIP: Session Initiation Protocol," RFC 2543, March 1999.

[6]   Fielding, R., J. Gettys, J. Mogul, H. Frystyk, L. Masinter, P. Leach, and T. Berners-Lee, "Hypertext Transfer Protocol — HTTP/1.1," RFC2616, June 1999.

[7]   Postal, J., "User Datagram Protocol," RFC 768, 1980.

[8]   "Transmission Control Protocol," RFC 793, 1981.

[9]   Stewart, R., Q. Xie, K. Morneault, C. Sharp, H. Schwarzbauer, T. Taylor, I. Rytina, M. Kalla, L. Zhang, and V. Paxson, "Stream Control Transmission Protocol," RFC 2960, October 2000.

[10]   Allen, C., and T. Dierks, "The TLS Protocol Version 1.0," RFC 2246, January 1999.

[11]   Rescorla, E., and N. Modadugu, "Datagram Transport Layer Security," RFC, November 2005.

[12]   The Third Generation Partnership Project, http://www.3gpp.org/.

[13]   Levin, O., A. Houri, E., and E. Aoki, "Inter-Domain Requirements for SIP/SIMPLE," IETF Internet-Draft, work in progress, February 2005.

[14]   Sparks, R., A. Johnston, and D. Petrie, "SIP Call Control—Transfer," IETF Internet-Draft, work in progress, February 2005.

[15]   Rosenberg, J., and H. Schulzrinne, "An Offer/Answer Model with the Session Description Protocol (SDP)," RFC 3264, June 2002.

[16]   "Packet-Based Multimedia Communications Systems," ITU-T Recommendation H.323v5, 2003.

[17]   Schulzrinne, H., and C. Agboh, "Session Initiation Protocol (SIP)-H.323 Interworking Requirements," RFC 4123, July 2005.

[18]   Andreasen, F., and B. Foster, "Media Gateway Control Protocol (MGCP) Version 1.0," RFC 3435, January 2003.

[19]   Cuervo, F., N. Greene, A. Rayhan, C. Huitema, B. Rosen, and J. Segers, "Megaco Protocol Version 1.0," RFC 3015, November 2000

[20]   Schulzrinne, H., S. Casner, R. Frederick, and V. Jacobson, "RTP: A Transport Protocol for Real-Time Applications," RFC3550, July 2003.

[21]   Schulzrinne, H., and S. Casner, "RTP Profile for Audio and Video Conferences with Minimal Control," RFC 3551, July 2003.

[22]   "Free Internet Telephony That Just Works," http://www.skype.com.

[23]   Berson, T., "Skype Security Evaluation," http://www.skype.com/security/files/2005-031% 20security%20evaluation.pdf.

[24]   Fabrice, D., "Skype Uncovered: Security Study of Skype," EADS-CCR http://www.ossir. org/windows/supports/2005/2005-11-07/EADS-CCR_Fabrice_Skype.pdf.

[25]   Johnston, A., "SIP, P2P, and Internet Communications," Internet-Draft, work in progress, March 2005.

[26]   Bryan, D., and C. Jennings, "A P2P Approach to SIP Registration," Internet-Draft, work in progress, January 2005.

[27]   Singh, K., and H. Schulzrinne, "Peer-to-Peer Internet Telephony Using SIP," Columbia University Technical Report CUCS-044-04, New York, October 2004.

[28]   Mahy, R., B. Biggs, and R. Dean, "The Session Initiation Protocol (SIP) Replaces Header," RFC 3891, September 2004.

4

Internet Threats and Attacks

4.1 Introduction

This chapter introduces and discusses a variety of common attack types and methods encountered on the Internet. While examples and applications of many kinds of attacks are discussed here, the purpose of this chapter is to familiarize readers with Internet attacks in general, and the attacks early adopters of VoIP systems have documented. Readers interested in a deeper understanding of Internet attacks should refer to additional sources that focus specifically with hacking, anti-hacking, network forensics, and incident response [1–4].

4.2 Attack Types

In this section, we will describe some common Internet attack types. Note that these attacks are not, in general, specific to VoIP and are encountered on the Internet for almost any application.

4.2.1 Denial of Service (DoS)

As the name suggests, the purpose of a denial of service (DoS) attack is to overload a computer (host, switch, server) or application and cause it to slow down or cease operating entirely. DoS attacks can be grouped into two classes: flooding and exploitation.

4.2.1.1   DoS Flooding Attacks

In a *flooding* attack, an attacker directs large volumes of traffic at a target or set of targets and attempts to exhaust resources of the target(s) such as bandwidth, CPU processing, memory, or even storage. Flooding is an apt analogy for this type of attack. Just as relentless waves of ocean water will eventually exceed the intended capacity of a levee and cause it to break, a similarly relentless flow of traffic exceeding the capacity of a switch, server, application or circuit will exceed that resource's capacity to operate, and it will fail. By causing the resource to fail, the attacker succeeds in denying or seriously retarding service to legitimate users of the resource.

A DoS flooding attack can be directed at different levels of the TCP/IP protocol stack. For example, directing a large volume of IP or Internet Control Message Protocol (ICMP) traffic at the target can be sufficient to cause congestion and wholesale packet discard, including discard of legitimate traffic. In a "smurf" attack, for example, an attacker sends ICMP echo request packets ("pings") to broadcast addresses of IP networks from a forged (spoofed) IP address. The host at the forged address (the target) is deluged with ICMP echo responses, and the network over which the IP broadcast is sent is often congested as well [5]. In this case, the nature and content of the packets do not matter, as long as they have the target as the destination IP address and the traffic volume is greater than the resources at or along the path to the target.

SYN flooding is a common TCP level attack. In a SYN flood attack, the attacker submits TCP connection requests (TCP SYN packets) faster than your server can process them. Your server must respond to an incoming TCP SYN packet with a SYN-AK packet and wait for a reply (TCP AK) to complete TCP's three-way handshake. If no reply is forthcoming, your server suspects that it's lost and retransmits the SYN-AK. The attacker knows this, so he withholds replies to your server's TCP SYN-AKs *and* continues to request additional TCP connections, expecting that your server will persist in retrying and keep the connection "half opened." Unless you configure your server to detect this hostile activity and adjust how it responds, the attacker will force your server into devoting its resources for connections that won't ever complete, leaving nothing for connections your users might wish to make.

Attackers can, in general, use IP, ICMP, and TCP level DoS attacks to attack any network equipment (VoIP endpoints, proxies, and servers) used to provide VoIP service that is not either configured to thwart such attacks, or protected by a security system capable of DoS prevention. DoS attacks against VoIP service are not limited to network and transport level attacks. At the application layer, the flooding traffic can be SIP messages. SIP message floods force the targeted SIP system to parse, process, and possibly generate responses. This activity consumes bandwidth as well as resources on the server and at intermediate systems (e.g. SIP proxies) along the signaling path. Cryptography does not prevent

DoS attacks. In fact, cryptography can make DOS attacks more effective. For example, VoIP systems that employ cryptographic authentication and message encryption can be flooded with forged messages. VoIP systems attacked in this manner can waste even more resources processing encrypted messages than plain text messages.

Following are several types of flooding attacks that use VoIP protocols.

*Control packet floods.*The attacker floods VoIP servers or endpoints with unauthenticated call control packets. If H.323 signaling is used, the attacker will send floods of GRQ, RRQ, or URQ packets. If SIP signaling is used, the attacker floods the target with INVITE, REGISTER or response messages. This form of DoS attack can easily be distributed. Any open administrative and maintenance port on call processing and VoIP-related servers can be a target for this attack. If successful, the attacker prevents the target device from processing new calls and may cause in-session calls to disconnect [6].

*Call data floods.*The attacker floods a VoIP endpoint or proxy with call data traffic attempting to exhaust that device's CPU or exceed bandwidth available to the endpoint. One form of such an attack is launched from a server, which typically has greater capacity than a VoIP endpoint. The attack uses SIP and RTSP to open large numbers of RTP sessions and push media at the targeted endpoint [7]. RTP sessions can consume considerable bandwidth, especially when uncompressed media {high quality video} are transferred, so this is a good example of a bandwidth exhaustion attack. If successful, the attacker prevents a VoIP endpoint from processing new calls and may cause in-session calls to disconnect. While the phone is under attack, the user cannot retrieve voicemail, access feature set configuration data, or call customer support.

*VoIP application flood attacks.*The attacker floods a voicemail or Short Messaging Service application (or server) with messages. The goal is to (a) keep the application busy and inhibit legitimate users from leaving or retrieving voicemail or text messages, (b) force the application to record unsolicited messages until all mailboxes have exceeded defined message limits, to prevent legitimate callers from leaving messages for subscribers, or (c) force the VoIP application to halt service because its storage capacity is exceeded.

An attacker can also attempt to deny service by attacking services that VoIP systems rely upon for correct TCP/IP operation. The theory behind such attacks is simple: if VoIP systems can't connect to IP networks, they can't place calls. By launching a DoS attack against servers that support Dynamic Host

Configuration Protocol (DHCP), DNS, or BOOTP, an attacker can inhibit and block not only voice but data applications as well. For example, in networks where VoIP and data endpoints rely on DHCP-assigned addresses, a DoS attack that disables a DHCP server prevents all endpoints from acquiring IP addresses, DNS and routing (e.g., default gateway) information.

4.2.1.2 Exploitation DoS Attacks

In an exploitation attack, the attacker tries to find some implementation flaw that will cause a target to fail. We consider this a denial of service attack because the service rendered by the targeted resource is unavailable until it is restored. Exploitation attacks do not always require floods of traffic. For example, a single, cleverly crafted URL can be sufficient to cause an Apache or Microsoft web server to terminate abnormally (crash). Some exploitation attacks try to force buffer overflow conditions (see Section 4.3.3); for example, the well-known SQL Slammer exploit forced a buffer overflow in the Microsoft SQL Server Resolution Service. Attackers commonly abuse URL encoding in HTTP or HTTPS requests to force a web application into doing something the designers didn't anticipate (e.g., cross-site scripting, SQL injection). Attackers are particularly interested in exploitation attacks that result in a privilege escalation, a situation where a web application's response to a malformed URL provides the attacker with administrative privileges [8].

Exploitation attacks are also known as implementation DoS attacks, and are not limited to web applications. An attacker can send malformed SIP, H.323, and RTP packets to VoIP servers to exploit a protocol implementation vulnerability, force a failure condition, and ideally, "get root" privileges (e.g., system administrator level access). Attackers have used malformed H.323 packets to exploit a Windows ISA memory leak and exhaust resources (CVE-2001-00546) and to exploit Nortel BCM DoS vulnerabilities [9]. Attackers have also used malformed SIP INVITE messages to deny service and execute arbitrary code on the Alcatel OmniPCX Enterprise 5.0 Lx (other SIP implementations were affected as well, see CERT Advisory CA-2003-06 [10].)

VoIP endpoint implementations (IP hard phones) are relatively new, and many devices, particularly wireless handheld VoIP phones, have limited CPU and memory compared to the average laptop or PC. These devices represent a veritable green field of opportunity for exploitation DoS attacks. Attackers have already demonstrated that some of these devices are vulnerable to fragmentation-based DoS attack methods (CAN-2002-0880), DHCP-based DoS attacks (CAN-2002-0835). VoIP servers and proxy applications will be susceptible to DoS exploits that target general purpose commercial operating systems that host such applications.

4.2.1.3   Distribute Denial of Service (DDoS) Attacks

Any host or server with limited bandwidth, storage, or resources is a possible target for DoS. However, applications can be designed and configured to provide some protection against DoS. For example, a SIP proxy server application can operate in a stateless mode for certain unauthenticated transactions. Server operating systems can also be configured to detect and prevent resource depletion, such as the SYN Attack Prevention in Windows Server 2003 [11].

The ubiquity of Internet access puts VoIP servers that might seem "over-provisioned" at risk to DoS attacks. Against such formidable targets, attackers employ distributed denial of service (DDoS) techniques. Over time, attackers can gain control of a large number of hosts, typically via the installation of a trojan program delivered as a virus (worm). Once installed, the trojan programs—"zombies"—are directed by an attacker's master program to simultaneously launch an attack against a target. While one attacking host may not suffice to cripple a well-provisioned server, thousands are much more likely to succeed.

A DDoS attack is illustrated in Figure 4.1.

Any protocol or application that can perform packet amplification can be used in a DoS attack. For example, a message exploder (which is truly as dangerous as it sounds unless proper security and consent mechanisms are implemented) can cause a single message to be sent to thousands of locations, or the same message sent thousands of times to a single location. Another would be to send a signaling request to establish a session with a high bandwidth streaming server (such as a video media server) with the media destination set to the host to be attacked. Completely secure and authorized signaling can be used in this

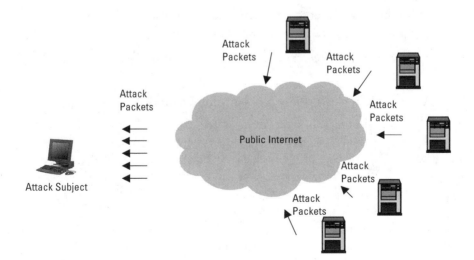

Figure 4.1   DDoS attack.

attack scenario if there is no corresponding authorization for the flow of media. A solution to this problem is discussed in Chapter 13.

Additional (external) DoS countermeasures can and should be provided as part of a layered defense. For example, modern firewall and intrusion prevention appliances detect and block many forms of IP, ICMP, and TCP DoS attacks. These security systems are typically deployed at a network perimeter, but they can also be placed in front of LAN segments that host VoIP servers (VoIP server farms) to protect against insider attacks [12]. Emerging, VoIP-aware firewalls and intrusion prevention systems detect and block VoIP-specific DoS attacks [13]. As products mature, advanced voice firewalls will detect and block mal-formed SIP and media packet headers; set, monitor and enable thresholds to prevent signaling-based flooding; and provide rate limiting to prevent band-width misuse.

Denial of service attacks and prevention are subjects of ongoing research. *Internet Denial of Service: Attack and Defense Mechanisms* [14] is an excellent resource for anyone who is interested in the principles and mechanics of DoS and DDoS attacks and methods to detect and prevent them.

4.2.2 Man-in-the-Middle

The man-in-the-middle (MitM) is the classic attack and many cryptographic systems are designed to protect themselves against it. The assumption is that the attacker has somehow managed to insert himself between the two hosts. As such, the attacker has the ability to:

- Inspect any packet between the two hosts;
- Modify any packet sent between the two hosts;
- Insert new packets sent to either hosts;
- Prevent any packets sent between the hosts from being received.

An example of a MitM attack is shown in Figure 4.2.

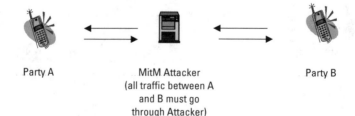

Party A   MitM Attacker   Party B
                     (all traffic between A
                      and B must go
                      through Attacker)

Figure 4.2 Man-in-the-middle attack.

A man-in-the-middle attack is a spoofing or impersonation attack, and can be executed at many levels of the TCP/IP architecture. For example, an attacker can use a laptop with a wireless LAN (WLAN) adapter, a means of signal amplification, and software to impersonate an access point (AP). Since WLAN stations seek out the most powerful signal, they will associate with the attacker's laptop rather than the real AP for the WLAN. By duping WLAN stations in this fashion, an attacker inserts himself in the middle of all connections made over the WLAN. Attackers can impersonate Ethernet hubs and poorly configured LAN switches in this manner as well.

An attacker can also execute a MitM attack at the application layer, for a specific protocol such as SIP. For example, by sending spoofed ARP packets on a LAN segment, an attacker can "poison" a host's ARP cache and draw traffic away from a legitimate SIP proxy server to his own impersonating server. By masquerading as a SIP proxy, the attacker can steal user account credentials; monitor, capture or modify signaling traffic; and redirect and forward calls placed through his proxy to other masquerading systems (e.g., a rogue voice mail server).

A MitM attacker who intercepts a SIP INVITE sent to its outbound proxy server can impersonate the proxy server and route requests maliciously. As another example, a compromised session border controller can be used to very effectively launch very damaging VoIP MitM attacks. MitM attacks can also utilize DNS poisoning techniques to draw users and VoIP endpoints away from legitimate servers to impersonating servers [15].

Cryptographic security measures can protect against certain MitM attacks. Authentication can prevent the attacker from posing as another host. For example, to thwart an attempt to impersonate a SIP proxy server, a UA can authenticate the proxy server before it attempts to place a call. Confidentiality can prevent the attacker from monitoring traffic. For example, the confidentiality provided by a secure tunneling protocol (IPSec or TLS) can provide protection against MitM. Integrity protection can detect when a MitM attacker is attempting to modify packets. For example, when SIP traffic is hashed and signed using a message authentication code, communicating UAs and proxy servers can detect attempts to alter signaling messages.

Many of the attacks described in the remainder of this chapter can be thought of as forms of the MitM attack, as many of them require that the attacker situate himself "between" communicating VoIP endpoints and servers.

4.2.3 Replay and Cut-and-Paste Attacks

A replay attack is one in which an attacker captures a valid packet sent between the hosts and resends it to some advantage to himself. Replay attacks are often used to impersonate an authorized user. If an attacker can successfully monitor and capture passwords used in clear text authentication schemes used in such protocols as PPP (PAP), POP3, and telnet, he can later replay the captured password when the PPP, POP3 or telnet server challenges the user for proof of identity. Replay attacks can be prevented by using one-time session tokens (one-time passwords, OTPs), multifactor authentication (e.g., SecurID), timestamps, and message authentication codes. An example replay attack is shown in Figure 4.3.

Capturing and resending packets at the application layer can cause a number of problems if replays are not detected. For example, if an attacker replays a VoIP signaling packet used to establish a PSTN call, he can place multiple phone calls when the caller only intended that one be placed, and the victim may incur excessive toll charges. If the attacker replays media packets, he can introduces discontinuities or dropouts in the audio stream.

Replay attacks can wreck havoc on multiparty calls. If RTP is used without authenticating RTCP packets and without sampling SSRCs, an attacker can inject RTCP packets into a multicast group, each with a different SSRC, and force the group size to grow exponentially [16].

A variant on a replay attack is the cut-and-paste attack. In this scenario, an attacker copies part of a captured packet with a generated packet. For example, a security credential can be copied from one request to another, resulting in a successful authorization without the attacker even discovering the user's password. Cut-and-paste attacks illustrate why credentials must be protected using measures that associate the credential to a particular request or transaction. In Chapters 9 and 10, we discuss how replay protection can be provided in SIP and SRTP.

4.2.4   Theft of Service

A theft of service attack can be launched in a number of ways. It can begin with the theft of a credential, shared secret key, or private key which can then be used

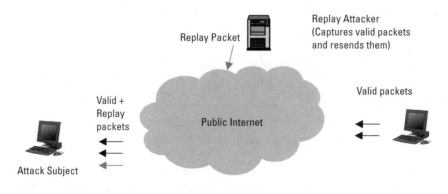

Figure 4.3 Replay attack.

to utilize resources without paying for them. Alternatively, theft of service can take the form of a redirection attack, in which the content (e.g., a voice message) or service requested by an authorized host or party can be redirected to an unauthorized host or party.

In a VoIP network, theft of service can be initiated for PSTN gateway access or use of a media relay resource. Other shared resources, for example, a video and audio conferencing bridge, can be targets of theft of service attacks.

4.2.5 Eavesdropping

Eavesdropping is a passive, and often difficult to detect, MitM attack in which the attacker copies or listens to communication between two hosts.

VoIP eavesdropping can be performed on signaling and media. Attackers eavesdrop signaling traffic to discover credentials, calling patterns, or identity or other sensitive information. Attackers eavesdrop media traffic to capture, replay or rebroadcast audio, video, or (text) messaging.

The danger of VoIP eavesdropping depends on the topology and underlying technology (switching systems and communications media) of the IP network used for voice transport. Eavesdropping on packets by tapping into fiber-optic circuits or by breaking into core switches that comprise eavesdropping on the Internet backbone is much more difficult than eavesdropping on traffic in a shared Ethernet segment on an unencrypted wireless link, or by breaking into an access point or broadband access router. Confidentiality measures protect against eavesdropping attacks.

One method to provide confidentiality is called hop-by-hop security. This method uses confidentiality measures on each network over which VoIP signaling and media traffic will be transmitted. For some network hops, confidentiality services specifically provided by underlying media can be used; for example, WiFi Protected Access (WPA™) or WPA2™ over wireless LAN in a company's branch office. For other network hops, higher level secure tunneling protocols can be used. For example, IPSec tunnels can be used to protect traffic from VoIP endpoints and proxy servers in a company's branch offices to an enterprise gateway located in the company's main office. In some cases, organizations may rely on isolation and trust, and not use encryption over an internal network hop. For example, a company may require all VoIP endpoints to establish a secure tunnel with TLS to connect to a SIP proxy server in its main office. Topology constraints imposed on the company's internal network may assure that traffic forwarded from the SIP proxy server to other VoIP servers and enterprise gateways uses LAN segments or VLANs with restricted connectivity (e.g., only voice servers may be connected to these segments).

Hop-by-hop security may be sufficient for some, perhaps many enterprise VoIP applications, but it has some noteworthy limitations, especially when individual users and organizations cannot exercise administrative control and implement security measures over every hop. As is the case with remote data access today, this is generally the case for mobile users, users (subscribers) of public VoIP services, such as users with guest credentials on a business partner's network. In such scenarios, the optimal way to provide not only confidentiality services but authentication and integrity as well is to encrypt traffic from user agent to user agent, or end-to-end.

4.2.6   Impersonation

Impersonation is described as a user or host pretending to be another user or host, especially one that the intended victim trusts. For example, in phishing attacks [17], an attacker tries to lure a targeted user to visit his web site instead of the user's online banking site, eBay account maintenance site, or company intranet portal. If the bait is successful, the attacker continues the deception at an impersonation web site to make the victim disclose his banking information, employee credentials, and other sensitive information.

Impersonation attacks often follow social engineering and replay attacks, using identities and credentials an attacker has "borrowed," stolen, eavesdropped, or otherwise managed to discover (such as by brute-force cracking of a weak password). The probability of success in impersonation attacks can depend on the way in which identity is being asserted and challenged. For example, organizations that use simple, static passwords for user accounts are generally more vulnerable to impersonation attacks than organizations using multi-factor authentication measures or digital certificates.

Address spoofing is a common form of impersonation attack. An address spoofing attack occurs when an attacker's device masquerades as the IP and/or MAC address of legitimate endpoint or server. By impersonating an address in this manner, the attacker causes all traffic to be redirected to his device.

4.2.7   Poisoning Attacks (DNS and ARP)

DNS poisoning targets the domain name service application. DNS is often used as the first step in establishing communication, to obtain the numeric IP address associated with a domain name or SIP URI or to resolve an ENUM query. An attack on the DNS can be used in replay, DoS, impersonation, and other attacks.

For example, SIP uses DNS to locate SIP servers for a given host name. If false DNS responses were generated in response to these queries, the SIP requests can be sent to a rogue proxy server.

As it is deployed today, DNS is vulnerable to several types of attacks, at different levels of the domain name system. An attacker can use a virus or spyware to modify the root server or default name servers a host uses to resolve domain names and direct the host to the attacker's name server, which will return incorrect and malicious responses to any DNS queries. An attacker can use a pharming attack to alter DNS responses that a host or local name server has chosen to store or cache locally, to avoid repeated referrals for a frequently requested domain name or URI. Attackers can even attempt to spoof name servers into transferring a zone file that contains altered DNS resource records.

To combat DNS poisoning, DNS caching can be disabled at endpoints and local resolvers; alternatively, resolvers can be configured to only cache resource records received from authoritative servers [18]. DNS name server operators can follow secure configuration, change and patch management "best practices" as recommended by security experts [19, 20] and organizations like SANS [21] to prevent poisoning and incorrect zone transfers.

Another way to combat this attack is to incorporate authentication and integrity measures in the DNS protocol itself such as DNSSec (DNS security, [22]). DNS security provides much-needed security services to DNS, including authentication of the origin of DNS data, data integrity protection, and authenticated denial of existence of DNS resource record sets. When DNSSec is deployed, all responses are digitally signed, allowing a DNS resolver to verify that the information it receives is identical to the information maintained by the authoritative DNS server. Private DNS root operators can deploy DNSSec today, but the hierarchical nature of the DNS makes widespread deployment challenging. For example, relying on digital signatures over one or a few levels of name delegation provides incomplete protection. For complete protection, the entire hierarchy must implement signing, beginning at the root of the DNS.

The address resolution protocol (ARP) is a request-response protocol IP-enabled devices use to discover the LAN (or WLAN) MAC address that corresponds to a specified IP address. If an attacker sends forged ARP packets to VoIP endpoints and server, these systems will modify their local forwarding tables and replace any previous table entry for this IP address with the attacker's device addresses. The ARP "cache" of these devices is now poisoned, and all future traffic to the target address will now be forwarded to attacker's device instead of intended device.

4.2.8 Credential and Identity Theft

A credential or identity theft attack provides an attacker with a valid identity and authentication credentials, such as a username and password, a shared secret, or a PIN that provides access to a private key. Any of these can then be used by the attacker to launch other attacks such as theft of service or impersonation.

Attackers can steal credentials over the wire by attacking the signaling protocol. Alternatively, attackers can use viruses, worms, and spyware to install key logging or trojan software that searches computer (and networked storage) for files containing account information. Increasingly, attackers are using phishing attacks. An attacker can place an incoming VoIP call with a forged identity to impersonate a service call and get the information using "social engineering."

Credential and identity theft is as much a social condition as it is a technical problem. We have identified many security measures that can be implemented to protect credential "data" while in transit and at rest. However, security measures alone, including stronger authentication techniques, cannot prevent users from unintentionally disclosing credentials, or prevent unsophisticated users from being deceived by a talented and financially motivated attacker. Disincentives and punitive actions against employees whose credentials are stolen have proven ineffective. Security awareness and public education are often a better investment for organizations seeking to minimize credential theft. Perhaps the most important defense against credential theft is providing strong incentives to employees and users to protect credentials.

4.2.9   Redirection/Hijacking

A redirection is a MitM attack in which one endpoint of a communication is maliciously changed, usually after identification, authentication, and authorization steps have taken place. For example, a VoIP signaling message may contain information about where the associated media should be delivered. If the attacker changes only the media delivery location information but leaves all other information unchanged, the resulting media can be redirected to another host. A redirection attack is shown in Figure 4.4.

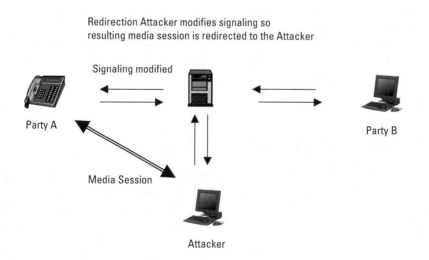

Figure 4.4   Redirection attack.

Integrity protection prevents such an attack, as any unauthorized changes/modifications to the communication can be detected. A man-in-the-middle can redirect or hijack a VoIP call by modifying the `Contact` header field of a SIP `REGISTER` request. If the modification is not detected, the `REGISTER` request will authenticate properly, but SIP requests will now be routed to the UA the attacker has identified as the `Contact`. Integrity protection measures protect against this form of attack.

Call redirection can be used to steal identity, credentials, or other sensitive information. If an attacker succeeds in forging a SIP Re-`INVITE` packet and convinces an endpoint to redirect an existing call to another device, such as a rogue voice application, the attacker can execute a voice equivalent of a phishing attack. Once the caller is redirected, the rogue voice application can try to deceive the caller into revealing a PIN or other personal identifying and sensitive information.

4.2.10 Session Disruption

Session disruption describes any attack that degrades or disrupts an existing signaling or media sessions or a session whose establishment is pending. For example, if an attacker is able to forge failure messages and inject them into the signaling path, he can cause sessions to fail when there is no legitimate reason why they should not continue. Similarly, if an attacker is able to inject disconnect messages, he can cause untimely or premature termination of media sessions. Alternatively, if an attacker introduces bogus packets into the media stream, he can disrupt packet sequence, impede media processing, and disrupt a session. Session disruption can be accomplished in SIP by injecting false signaling packets. For example, when HTTP Digest authentication is being used, an attacker can inject malformed `ACK`s or `CANCEL`s to cause calls to fail during setup.

An example of session disruption is shown in Figure 4.5.

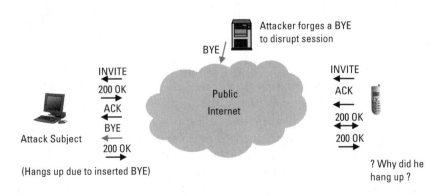

Figure 4.5 Session disruption attack.

Several types of disruption attacks have been demonstrated, including the following:

Delay attacks are those in which an attacker can capture and resend RTP SSRC packets out of sequence to a VoIP endpoint to force the endpoint to waste processing cycles re-sequencing packets and degrade call quality on CPU-challenged devices.

Tear-down attacks allow an attacker to send forged RTP SSRC collision and RTCP BYE to signal a failure condition or force call disconnection [23].

QoS Modification enables an attacker to alter quality of service information used by underlying protocols to increase latency and jitter. By altering IEEE 802.1Q VLAN tags or IP packet ToS bits, the attacker could disrupt the quality of service engineered for a VoIP network. This attack can be executed as a man-in-the-middle or by modifying a VoIP endpoint device configuration.

Disruption of underlying media allows an attacker to disrupt service provided by IEEE 802.1x-enabled networks by spoofing frames such as EAP-Failure to force endpoints to disconnect from the network. An attacker can disrupt Voice over WLAN service by disrupting IEEE 802.11 WLAN service using, for example, radio spectrum jamming or a WPA MIC attack. (This latter attack exploits a standard WLAN security measure: a wireless access point will disassociate stations when it receives two invalid frames within 60 seconds, causing loss of network connectivity for 60 seconds. A one-minute loss of service is hardly tolerable in a voice application.)

4.3 Attack Methods

In the following sections, a number of attack methods will be discussed.

4.3.1 Port Scans

Port scanning is a technique in which an attacker tries to open connections to listening ports on a server. If a particular, well-known port accepts a TCP connection, the attacker can attempt attacks specifically designed to exploit the protocol and application that is likely to be executing on the server. VoIP servers typically use well-known ports, such as port 5060 for SIP. Ports commonly scanned as potential attack vectors include HTTP (80), SMTP (25), Sequel server (SQL, 1433), and DNS (53). Port scanning is so widespread that security administrators can find it hard to determine when their systems are being intentionally targeted.

4.3.2 Malicious Code

Malicious code is an umbrella term that describes any form of software that is installed without authorization, notice, or consent from the computer owner or administrator. Once installed, the activities the malicious code performs are often used to classify the code as virus, worm, spyware, trojan program, or blended threat.

VoIP clients and servers that run on general purpose computing hardware are as susceptible to malicious code attacks as computers used for data applications. Best practices for securing VoIP endpoints and servers thus include secure configuration (hardening), disabling of non-essential services, careful administration of user accounts, timely and complete installation of operating system and application patches and software updates. Complementary security software such as antivirus, antispyware, host intrusion detection and software firewalls on VoIP systems play an important role in securing VoIP services.

Handheld VoIP endpoints are particularly vulnerable to malicious code attacks. Many of these devices run on lightweight operating systems (Windows CE, Palm OS, SymbianOS, and POSIX), and are not as likely to have antimalware software {e.g., antivirus, antispyware, and antispam). Malicious code may render the device unusable, or, as in the case of mass mailing worms (see Section 4.3.2.2), the malicious code may attempt to spread itself or deliver unsolicited messages (see SPIT, Chapter 13). Worm-like activity can be particularly problematic for both subscriber and VoIP operators if calls placed by the worm incur usage charges.

4.3.2.1 Viruses

A virus is malicious code that attacks or takes control of a host. A virus needs a transport for it to be spread from host to host and replicate itself. Various communication protocols can be used as transports for viruses. A virus that has its own transport built in is known as a worm.

Viruses can be spread by email attachments, web pages, instant messages, program macros, hidden files in zipped archives, and many other techniques All can be extremely destructive. Once installed, viruses can erase data files, modify computer settings, or remove critical system files to cripple the host computer.

4.3.2.2 Worms

A worm is a piece of malicious code that is commonly delivered via e-mail. Once a worm installs itself on a host, it seeks to replicate by installing itself on other hosts.

4.3.2.3 Trojans

A trojan is a piece of computer code that masquerades as something innocuous or beneficial, but, in fact, is used to spread viruses, install other malicious programs such as keyloggers, DDoS zombies, remote administration toolkits (also known as root kits), e-mail and web and peer-to-peer servers. Trojan programs may modify local computer configuration settings, disable security measures and event logging to hide their presence. Servers installed by trojan programs can be used to send spam, host illicit web sites and chat rooms, and offer illegal downloads of copyrighted software and music.

4.3.2.4 Spyware

Spyware is similar to a trojan program in that it is installed, without permission, on a host, and then tries to remain undetected. Different kinds of spyware behave differently. Tracking and mining spyware gathers web browsing and application usage and reports this information back to an ad server over the Internet. This information may be used to feed another spyware component on the infested computer with targeted advertising. Spyware writers can also include keyloggers to obtain passwords, credit card information, and authentication credentials. Various forms of spyware alter browser favorites settings and replace default search engine settings with biased search engines to drive traffic from sites the user intends to visit to a spyware company's affiliate site. Some modify the browser home page to visit a gambling or pornography web site when the user launches a browser. Other spyware even modifies a computer's TCP/IP settings, altering where the PC resolves domain names.

Spyware has quickly become a pandemic on the Internet, and there is every reason to believe that spyware variants will be developed to exploit VoIP systems. Spyware installed on a VoIP phone might monitor a user's calling patterns, autodial calls to an affiliate merchant's salesperson, change a user's default directory assistance number to an affiliate directory provider that charges exorbitant fees, or direct a user's voice mail to a server that can literally hold the message hostage. There's also no reason to believe that spyware won't be used to gather personal and credit card information from VoIP systems.

4.3.2.5 Blended Threat

A blended threat combines elements of viruses, worms, trojan programs, and malicious code designed to exploit known server and Internet vulnerabilities to initiate, transmit, and spread an attack A blended threat propagates even more quickly than mass mailing worms by using any propagation path it can find on the computers it infects (e.g., by exploiting common network services such as file sharing, ftp, telnet, and even VPN connections). A blended threat can work its way onto computers as an e-mail attachment. Once it gains administrative control of the infected computer, it may try to disable the computer's antivirus software and software firewalls and erase event logs [24].

4.3.3 Buffer Overflow

A buffer overflow is an attack technique that targets an input parameter or value in a program to deliberately cause a software exception condition (failure). For example, in a VoIP signaling protocol, the presence of variable length parameter fields can be exploited for a buffer overflow attack. If a field normally 10 to 20 characters long is fed a 200-character string, there is a possibility that a badly written piece of software may not handle this properly, and may crash.

In SIP, a well-written message parser is the best protection against buffer overflow attacks. In H.323, a reliable ASN.1 decoder will protect against a buffer overflow attack.

The University of Oulu in Finland has developed a test suite for both SIP and H.323 systems to test against various types of buffer overflow attacks. The tool is called PROTOS and has a SIP [25] and H.323 [26]. The test suite generates INVITE requests based on the SIP Torture Test draft [27] which is designed to exercise a parser. As an example of a sample torture test message:

```
INVITE sip:vivekg@chair-dnrc.example.com;unknownparam
SIP/2.0
TO:
sip:vivekg@chair-dnrc.example.com; tag = 1918181833n
from: "J Rosenberg \\\""<sip:jdrosen@example.com>
  ;
  tag = 98asjd8
MaX-fOrWaRdS: 0068
Call-ID: wsinv.ndaksdj@192.0.2.1
Content-Length    : 150
cseq: 0009
INVITE
Via  : SIP  /   2.0
/UDP
  192.0.2.2;branch=390skdjuw
s  :
NewFangledHeader:   newfangled value
continued newfangled value
UnknownHeaderWithUnusualValue: ;;,,;;,;
Content-Type: application/sdp
Route:
  <sip:services.example.com;lr;unknownwith=
  value;unknown-no-value>
v:  SIP  / 2.0  / TCP    spindle.example.com    ;
  branch  =   z9hG4bK9ikj8  ,
  SIP  /   2.0  / UDP  192.168.255.111  ; branch=
```

```
    z9hG4bK30239
m:"Quoted string \"\"" <sip:jdrosen@example.com> ;
newparam =
        newvalue ;
  secondparam ; q = 0.33

v=0
o=mhandley 29739 7272939 IN IP4 192.0.2.3
s=-
c=IN IP4 192.0.2.4
t=0 0
m=audio 49217 RTP/AVP 0 12
m=video 3227 RTP/AVP 31
a=rtpmap:31 LPC
```

While this test message actually is a completely valid SIP INVITE request, it has many unusual properties such as:

- Lots of line folding;
- An empty header field;
- Extra linear white space (LWS);
- Unknown header fields;
- Unusual header field ordering;
- Parameter with no value.

In addition to this message, the document includes requests with extra long fields. A good SIP parser will be able to handle these messages.

However, results from 2003 publicized by CERT [28] in "CERT® Advisory CA-2003-06 Multiple Vulnerabilities in Implementations of the Session Initiation Protocol (SIP)" [29] indicate that many commercial SIP implementations failed with these test messages. Some even crashed and went into a state that an attacker can use to compromise a system. The CERT Coordination Center (CERT/CC) is a not-for-profit associated with Carnegie Mellon University that publishes computer security incidents and recommendations.

A similar warning " CERT® Advisory CA-2004-01 Multiple H.323 Message Vulnerabilities" [30] was also issued by CERT in 2004 for some H.323 systems.

CERT, SANS, MITRE, and other vulnerability reporting and "bug tracking" organizations are extremely active in tracking down and identifying implementation flaws. These efforts help raise awareness among implementers that handling unexpected conditions and malformed packets in their programs is imperative.

4.3.5 Password Theft/Guessing

If a password is chosen by a human, it will not likely be very random. In cryptographic terms, the password has very low entropy, and is likely to be susceptible to a brute-force or dictionary attack in which commonly used words and word combinations are tried repeatedly until a match is found. Organizations can impose maximum lifetimes on passwords, enforce composition and complexity criteria, and limit re-use of passwords to reduce the possibility of successful dictionary attacks. Long, difficult to attack passwords can be generated by a service provider or enterprise, or user chosen passwords validated and rejected until they have a certain level of complexity.

When passwords are used, authentication measures should limit the number of repeated failed attempts. Instead, after a certain number of failures, subsequent attempts should be silently discarded for a period of time.

4.3.6 Tunneling

Tunneling is an attack method by which one protocol rides inside another, often for the purpose of traversing a firewall or other perimeter defenses. Tunneling is extremely common. Port 80, the well-known port for HTTP, is reused for a variety of protocols to pass through firewalls.

It is even possible to tunnel other protocols on top of VoIP protocols. For example, almost any arbitrary protocol can be tunneled on top of a VoIP signaling protocol such as SIP. SIP can also be used to connect a host to the PSTN, effectively acting as a modem on a device which may not otherwise have PSTN connectivity and reachability.

Good protocol and architecture design does not make use of tunneling. Deep packet filtering in firewalls and other content filtering approaches can block tunneling.

4.3.7 Bid Down

A bid down attack is one in which a man-in-the-middle interferes with the negotiation of a security parameter and succeeds in forcing the other parties to agree to a lower level of security. For example, if a VoIP session initiation gives the option of either an encrypted or unencrypted media session, a bid down attack would result in the selection of the unencrypted session even if both sides would prefer to establish the encrypted session.

Another example is if a MitM attacker modifies a HTTP Digest challenge and changes it to a HTTP basic challenge. If the UA responds with a HTTP Basic response, the UA has given the attacker the user's password in clear text. The protection against this is to only use a single authentication or security option and disallow insecure options. SIP UAs should not respond to a HTTP Basic challenge, which would cause this bid down attack to fail.

4.4 Summary

The security techniques discussed in this and subsequent chapters can help protect against many of these attacks. However, some attacks, such as DDoS and identity theft are extremely hard to prevent. Finally, VoIP security planning must be integrated with an overall Internet access security plan, as many of these attacks are not unique to VoIP.

References

[1]   McClure, S., Kurtz, G., and J. Scambray, *Hacking Exposed,* 5th Ed., Osborne/ McGraw-Hill, 2005.

[2]   Spitzner, L., *Know Your Enemy: Revealing the Security Tools, Tactics, and Motives of the Black Hat Community,* Reading, MA: Addison-Wesley, 2002.

[3]   Kruse, W.G. II., and J.G. Heiser, *Computer Forensics: Incident Response Essentials,* Reading, MA: Addison-Wesley, 2002.

[4]   Mandia, K., and C. Procise, *Incident Response: Investigating Computer Crime,* New York: Osborne/McGraw-Hill, 2005.

[5]   CERT® Advisory CA-1998-01, "Smurf IP Denial-of-Service Attacks," http://www.cert. org/advisories/CA-1998-01.html.

[6]   "Overlooked: PBX and Voice Security in a Networked World Herrera," http://www.giac. org/certified_professionals/practicals/gsec/2436.php.

[7]   Rosenberg, J., "The RTP DOS Attack and its Prevention," http://www.jdrosen.net/ papers/draft-rosenberg-mmusic-rtp-denialofservice-00.html.

[8]   Ollman, G., "URL Encoded Attacks," http://www.cgisecurity.com/lib/URLEmbedded Attacks.html.

[9]   CERT® Advisory CA-2003-06 "Multiple Vulnerabilities in Implementations of the Session Initiation Protocol (SIP)" http://www.cert.org/advisories/CA-2003-06.html#vendors.

[10] The Common Vulnerabilities & Exposures List, maintained by The MITRE Organization, http://www.cve.mitre.org/. (CVEs identified for SIP are found at http://cve.mitre. org/cgi-bin/cvekey.cgi?keyword=SIP, CVEs identified for H.323 are found at http://cve. mitre.org/cgi-bin/cvekey.cgi?keyword=H.323, etc.)

[11] "How to Harden the TCP/IP Stack against Denial of Service Attacks in Windows Server," 2003, http://support.microsoft.com/default.aspx?scid=kb;en-us;324270.

[12] "Protecting Windows Servers Against DoS Attacks," http://www.corecom.com/external/livesecurity/windowsdosattacks.htm.

[13] Collier, M., "The Value of SIP Security," http://www.networkingpipeline.com/security/ 22104067.

[14]   Mirkovic, J., S. Dietrich, D. Dittrich, and P. Reiher. *Internet Denial of Service: Attack and Defense Mechanisms.* Upper Saddle River, NJ: Prentice Hall, 2004.

[15]   Ollman, G., "The Pharming Guide," http://www.ngssoftware.com/papers/The Pharming Guide.pdf.

[16]   Rosenberg, J., and H. Shulzrinne, RFC 2762, "Sampling of the Group Membership in RTP," http://www.ietf.org/rfc/rfc2762.text.

[17]   Piscitello, D., "Anatomy of a Phishing Attack," http://hhi.corecom.com/phishingexpedition.htm.

[18]   Piscitello, D., "DNS Pharming: Someone's Poisoned the Water Hole!" http://www.corecom.com/external/livesecurity/dnspharming.htm.

[19]   Liu, C., http://www.linuxsecurity.com/resource_files/server_security/securing_an_internet_name_server.pdf

[20]   Snyder, J., "Domain Name System (DNS) Configuration, Management and Troubleshooting," http://www.opus1.com/www/presentations/sanug/index.htm.

[21]   SANS "DNS Poisoning Summary," http://isc.sans.org/presentations/ dnspoisoning.php, March 2005

[22]   DNS Security Extensions, http://www.dnssec.net/.

[23]   Collier, M., "Basic Vulnerability Issues for SIP Security," http://download.securelogix.com/librarydownload.htm?downloadfilename=SIP_Security030105.pdf.

[24]   Piscitello, D., "What Is a Blended Threat?" http://hhi.corecom.com/blendedthreat.htm.

[25]   University of Oulu, Oulu, Finland Electrical and Information Engineering.
       PROTOS SIP Test-Suite, http://www.ee.oulu.fi/research/ouspg/protos/testing/c07/sip/.

[26]   University of Oulu, Oulu, Finland Electrical and Information Engineering, PROTOS H.323 Test-Suite, http://www.ee.oulu.fi/research/ouspg/protos/testing/c07/h2250v4/.

[27]   Sparks, R., A. Hawrlyshen, A. Johnston, J. Rosenberg, and H. Schulzrinne, "Session Initiation Protocol Torture Test Messages," IETF Internet-Draft, work in progress, April 2005.

[28]   Computer Emergency Response Team (CERT), http://www.cert.org.

[29]   CERT® Advisory CA-2003-06) "Multiple Vulnerabilities in Implementations of the Session Initiation Protocol (SIP)," http://www.cert.org/advisories/CA-2003-06.html, February 21, 2003.

[30]   CERT® Advisory CA-2004-01, "Multiple H.323 Message Vulnerabilities," http://www.cert.org/advisories/CA-2004-01.html, January 14, 2004.

5

Internet Security Architectures

5.1 Introduction

This chapter provides readers who are unfamiliar with Internet network security terminology a good foundation in security technologies, cryptography, and best practices and standards for secure networking. Readers who are familiar with networking and security principles and basics may skip this chapter. Readers who do not have at least a modest understanding of Internet security architectures, services, implementation and industry best practices should find this chapter useful.

5.1.1 Origins of Internet Security Terminology

Internet security is often described in military terms. We can trace the origins of many of these terms to the castle building vocabulary of England during the reign of Edward II [1]. Castles were designed to protect items of value (property) and people of importance (the nobility, landowner and merchant classes) from miscreants, robbers, and armies of rival lords who would steal or destroy valuables, and injure the nobility, if not prevented from doing so. Castles began as single buildings, fortified towers called donjons or keeps. To further secure castles from attack, builders constructed layers of security to protect the buildings, their occupants and treasure. Initially these fortifications were rudely constructed from dirt, evolving to wooden stockades (the mott-and-bailey) and later, incorporating stone and additional defenses to become what many of us imagine when we hear the word "castle." We envision a formidable fortress surrounded by moats, accessible only via a drawbridge, with yeomen and archers positioned on formidable walls of stone, sworn by oath to keep intruders at bay.

73

Within these layers of defense, men at arms stationed at checkpoints allowed recognized inhabitants and authorized visitors to come and go within the confines of the castle walls, but only permitted a privileged few access to the castle itself. If the castle was attacked, alarm fires and bells were used to raise a general call to arms announcing that defenses had been breached. Bridges were drawn, gates were barricaded, and tripwires and mantraps were used to block and delay any intruders who managed to make their way past the initial lines of defense.

During this period, lords and their advisors decided what defenses were needed to protect the castle and what cost was necessary to do so. They decided which of the lord's holdings, valuables, and citizens would be protected within the confines of the castle and passed laws establishing who would be allowed admission. They chose military leaders, recruited men at arms, and hired mercenaries to defend the castle, and routinely inspected their armed forces and defenses. In short, it can be said that they set security policy, performed risk, threat and vulnerability assessment, and supervised security implementation. We discuss these concepts in Sections 5.2, 5.3, and 5.4, respectively.

The analogy between the defense of castles in the Edwardian period and network security is interesting, accurate, and powerful. These practices are familiar to anyone who is currently involved in Internet security definition, design, and implementation.

5.1.2 Castle Building in the Virtual World

Today, an enterprise relies on chief security officers, security professionals and qualified security staff to identify assets, identify threats, assess risk, and develop a security policy for the organization. They rely on these people to implement and maintain a strong security profile through a combination of physical and virtual world security measures. Today, security is needed by all Internet users, not just nobles and governments.

We deploy similar physical security measures to protect networked computing facilities, that is, Internet data and operations centers. We build a secure perimeter, a continuous fortification around our networks to protect our physical and electronic assets. Physical security measures to protect networks and communications systems still include walls and armed guards at checkpoints. Electronic sensors, laser tripwires and even modern incarnations of the mantrap are employed where the value of electronic assets and the systems on which they are stored or operated is particularly high (as in financial institutions). Electronic swipe cards and biometric devices (fingerprint, iris, and palm scans, and facial recognition) replace and complement armed guards as preferred methods of verifying identities of those who have authorized access to secure facilities.

Physical security alone will not cover the problems associated with protecting electronic valuables and trusted communities of individuals (insiders) from miscreants, competitors, terrorists and rogue governments (outsiders), who can access these assets via an organization's Internet connection(s) unless preventive measures are taken. Even so, additional security measures are often required. Descriptions of some of these measures follow:

Authentication systems provide the means to distinguish authorized users and systems (members of the trusted community) from unauthorized ones. We introduce authentication in Section 5.5 and consider the topic in some detail in Chapter 8.

Authorization services control access to resources based on permissions granted to authenticated identities in a security policy. We introduce authorization in Section 5.6.

Auditing and monitoring services document activities performed by individuals and systems which can be used to identify normal versus suspicious activities, to determine whether security implementation complies with intended security policy, and to respond to and assess the consequences of security incidents. We introduce auditing, monitoring, and logging in Sections 5.7 and 5.8.

Perimeter security and policy enforcement systems (packet-filtering, stateful inspection firewalls, application proxies, session border controllers) prevent unauthorized access and block attacks. We introduce these concepts in Section 5.9.

Intrusion prevention, detection, and blocking systems (intrusion detection systems (IDS), tripwires, honeypots, antivirus, and server integrity software and hardware) provide additional lines of defense within the secured perimeter, and provide alarms warning administrators of security breaches. We introduce these concepts in Section 5.10.

Policy-based network admission and endpoint control prevent devices that are judged "unsafe" from connecting to networks. We introduce admission and endpoint control in Chapter 7.

5.2 Security Policy

Every organization needs a set of guidelines that identifies the assets the organization deems valuable or sensitive, describes appropriate use and handling of such information, and defines a set of actions that constitutes authorized use and access, and how authorized parties are authenticated. A security policy identifies threats to an organization's assets and measures to take to mitigate or reduce these threats. A security policy documents processes for maintaining security and for responding to attacks or incidents. It identifies conditions that necessitate escalation, disclosure, and notification of law enforcement, public relations, and legal responses. A security policy says, "Here is what we value, here are the threats that put what we value at risk, here is how we intend to protect what we value, and here is what we will do if any asset should ever be lost, damaged, disclosed without authorization, stolen, or attacked."

Few security activities are as routinely overlooked and discounted as security policy development. A common complaint every security officer and professional hears is, "Why bother developing a security policy? The assets and technology of a business change constantly!" To this, experienced security professionals answer, "Change is a constant, not an excuse." Without guidelines on which to base security measures, organizations lack a comprehensive strategy for protecting assets, and so make ad hoc and technology-driven decisions. The result is an implementation where security measures cannot satisfy all the criteria for adequately protecting assets, and where too few members of an organization will know, appreciate, and consequently, comply with security and acceptable use policies (AUPs).

Why are security policies important? Lacking policies, an organization cannot know exactly what it is trying to protect, what measures the organization should implement to protect it, which parties are responsible for implementing these measures, and which parties are responsible for assuring that the measures are implemented as intended. The organization at large may not know how to react to an attack, and will have little or no basis to hold insiders or attackers accountable for their actions. When such action or response plans are not provided, maintaining security standards, responding to security breaches, and taking remedial action are ripe with uncertainties and become increasingly challenging as the organization grows.

Who should develop a security policy? Technical staff are important contributors, but when security policy is placed entirely in the hands of IT, the contributions "nontechnical" members of the organization can make to a security policy are overlooked and lost. The most effective security policies can be developed by engaging representatives from legal, accounting, auditing, human resources, and employees from all business units. These parties can provide valuable input, especially in circumstances where a security policy must consider an organization's liability, accountability, regulatory obligations, and business objectives. When a security policy is completed, it should be shared with all parties, who should sign on or otherwise acknowledge their responsibility and accountability to the policy, and then consider implementation.

When should a security policy be developed? Few organizations today have the luxury of developing a policy before computers are used and networked. This matters only with respect to how the security expressed in a policy is ultimately implemented. While we don't dismiss the challenge of incorporating security as a retrofit or upgrade to design rather than a fundamental consideration, it's still "an implementation." An important consideration regarding "when" is that a security policy should remain a living document within the organization. It should be revisited frequently, as assets, business objectives, and the regulatory environments that affect an organization change. Again, change is a constant, and security policy (and implementation) must respond and adapt to change.

Many resources, templates, and standards for developing a security policy are available on the Internet [2–5]. None may precisely match your organization's needs, but these can serve as guidelines.

5.3 Risk, Threat, and Vulnerability Assessment

During a risk assessment, an organization identifies and assigns a dollar value to its (electronic) assets. Assets may include materials for which you have copyrights and patents, intellectual property, research, databases containing accounting, personal or similarly sensitive or regulated material, strategic and marketing plans. Anything connected to a network (and Internet) that would cause you financial harm if lost, altered, replayed, or damaged, expose you to embarrassment or legal action, impair your ability to operate your business, or place you in violation of a government regulation is an asset.

Some assets are more tangible than others. The availability of an organization's voice and data networks, servers, and endpoint devices is an asset. Availability is a critical asset for any organization that relies heavily on voice communications and maintains a virtual merchant web presence, conducts a meaningful part of its business or derives substantial revenue online. Actually, loss of availability can be viewed as a liability. For example, every minute a merchant cannot place and receive calls, or cannot process transactions and sell goods represents lost income. In addition, companies whose communications and distributed processing for a manufacturing production line or parcel delivery service that could be delayed or blocked by an attacker are as much at risk for a loss that is related to availability as are e-merchants.

Once an organization identifies assets and assesses their values, it must identify the risks to attack or threats to which each asset is exposed. Unintended or malicious disclosure of pharmaceutical research, formulae, or testing is a serious threat to a drug company. Disclosure of a patient's medical information is a serious threat to a healthcare provider, as well as a violation of U.S. federal law. Denials of service are threats to any business that relies on voice and data applications. Modification of data is another kind of threat. Consider how changing the 4th or 5th decimal place of an interest rate used in amortizations will affect interest accrued or paid out. Imagine how a change in the 8th decimal place of the value of pi might affect calculations critical to an unmanned space probe.

When developing a security policy, it's helpful to create a table or matrix. Identify:

- Assets, their value, and location;
- The nature of threats to each asset (disclosure, modification, theft, loss);
- The ways in which assets are vulnerable to attack. These may include, for example:
  - Unintended or inappropriate disclosure of passwords, identity cards, and other personal information, leading to unauthorized system or facility access.
  - Unauthorized file access, leading to unauthorized disclosure or destruction of sensitive information.
  - Information integrity compromise, leading to errors in calculations, production problems, and hazardous conditions, among others.
  - Unauthorized access, leading to misuse, inappropriate use, or malicious operation.
  - Denial of service or forced system or (server) software failure, leading to service interruptions, loss of revenue and business communications.
  - Theft of equipment, possibly leading to many of the above attacks.
- Methods you intend to use for removing, mitigating, removing, or minimizing the threat (i.e., file permissions and encryption measures you will take to protect file accuracy and availability; how you will prevent unauthorized access; and how you will detect and block a denial of service attack).

Why is such a table important? If you don't catalog every asset and the spectrum of attacks that threaten your assets, how can you take measures to protect them? If you don't know how much is at stake should an asset be lost, destroyed, stolen, or disclosed to unauthorized parties, how can you determine if you're spending enough or too much to protect them? Similarly, if you don't know the approximate per minute or hourly value of offering e-merchant service, how can you determine if you are taking adequate measures to remain on line?

Risk, threat, and vulnerability assessment can be a very complex process, and it's important that you have credible valuation of assets should you seek financial restitution in courts of law following an attack. Hire certified outside security auditors to perform a risk, threat and vulnerability assessment. If you choose to perform these assessments yourself, review guidelines established by organizations such as The Information Systems Audit and Control Association & Foundation [6] before you begin.

5.4 Implementing Security

Implementing security begins with people, not technology. Security staff must understand networking concepts, Internet protocols, operating system and application architectures at a unique level. They not only must know how hardware and software systems are supposed to work, but they should have an appreciation for the ways they can be made to malfunction and how malfunctions can be exploited. They must be technically versed in the function, configuration, and operation of systems and services they will be responsible for securing, including operating systems, Internet servers (voice, mail, domain name, web), routers, switches, call managers, and security appliances. Finally, they must remain current with vulnerabilities and exploits of these systems.

More than any of these skills, however, security staff must be trustworthy. Thomas Wadlow suggests that good security experts can think evil thoughts but exercise restraint and good judgment not to act on them (see [7], p. 59). The keywords here are restraint and good judgment. Hiring "former hackers" may not be wise, as people who admit to having attacked systems have already demonstrated that they can't satisfy these criteria [8]. Since organizations hire people to whom they must entrust their assets, it is essential that they interview carefully, perform background checks appropriate to the job description, and "trust wisely" ([7], p. 48).

Implementing security is also more about processes, documentation, and discipline than technology. A discrete process—that is, a sequence of actions to be performed, and systems affected, including documentation (e.g., an audit trail)—should be defined for every security operation, and security staff must adhere to the process. Effective authentication, for example, is achieved by having a well-defined process for managing user identities and credentials. Yet even the best authentication technology will not compensate for poor account maintenance and lax policies for credential revocation, or employees who disclose passwords for Cadbury bars [9].

Security technology is an unfortunate necessity, and so a rigorous approach to assessing technology is important. Before making any investment in security technology and services, security staff should perform an assessment of security services and facilities available on each element of the network that is either treated as—or hosting—an asset identified in the security policy. Some questions security staff ought to consider include:

- What measures required by the security policy could these elements provide directly?

- Do they provide adequate monitoring and logging features, and are they easily audited?

- What complementing security technology and software are required to eliminate or reduce the threats identified?

- What levels of scale and availability are required from server and security systems?

- What criteria can be used to distinguish the best of breed among several competing and equivalent technologies (e.g., firewalls, VPN appliances)?

- How manageable is the technology (security is complex enough without having to deal with very complicated user interfaces)? Has it been subjected to and withstood reasonable scrutiny by the security community? How often are bugs reported? Is the vendor timely and responsive in patching reported vulnerabilities? Disclosure policy?

- What logging and monitoring features do network elements provide? Can log information be easily processed by third-party analysis software?

Additional questions and issues particular to a given security policy may also apply.

Implementing policy at the individual systems expected to provide security measures, called configuration, is a process that requires familiarity with how systems and network protocols, and user interfaces operate. Configuration should be subjected to internal review to confirm the configuration does in fact implement the policy intended. Where possible, a security audit by independent third parties is desirable. Configurations should be archived along with sensitive data as they are, in fact, some of the most sensitive data in the organization.

User interfaces are not always as intuitive and simple as one would hope. Complexity and ambiguity are more of a threat when implementing security measures than anywhere else, so security staff should know, and make use of, technical points of contact for vendors of their security systems and software. Security staff should complement knowledge of security measures they know with training and education on new practices, countermeasures, and technologies. No security staff member should represent a single point of responsibility and knowledge, and hence, a failure. Staff employment, like equipment deployment, should consider redundancy and diversity.

5.5 Authentication

Authentication is a process where an entity, either a user or computer system, confirms its, his, or her identity by presenting credentials that are difficult for anyone but the actual identity to produce. For user authentication, the credentials must provide adequate evidence that the user is who he or she claims to be. Airport security provides a familiar example of user authentication. An individual must present credentials that include a photograph and sufficient and accurate personal information to assure airline personnel that the credential holder and the ticketed passenger are the same person. In most cases, a passport or driver's license satisfies the credentialing criteria. For user authentication in network security, forms of credentials include passwords (something only this individual knows), hardware tokens (something only this individual owns), or a biometric (some physical characteristic, such as a voice imprint, fingerprint, iris or facial scan, even a DNA sample) that distinguishes this individual from all others.

For system authentication, digital credentials—for example, the private key of a certified public/private key pair, or the signed hash of a shared secret key—offer equivalent confirmation of this system's identity.

Authentication is the enabling process for all security. Without confirming the identity of an entity, no access and authorization should be granted to that entity. However, once an identity is confirmed, security policies associated with that entity could be used to determine what actions the entity may take. Just as a physician's identity must be authenticated before she is granted access to patient records in a hospital database, similarly, firewalls or VPN appliances must exchange and authenticate their identities before they establish a VPN connection.

Many authentication methods are used today. Some commonly used methods are:

- Plain text passwords (or pass phrases) are an alphanumeric sequence or phrase associated with a user identity. Passwords are notoriously easy to intercept and replay, or constructed to make them easy to remember and hence, easy to crack. Therefore, if they are used over untrusted networks, they must be transmitted over an encrypted tunnel.

- Trusted hostname systems, such as those used by UNIX "r" commands (rlogin, rsh, rcp), verify an identity based on the account and IP address of the host from which the user attempts to log in. This authentication method is as vulnerable to misuse as passwords.

- Temporal key distribution methods, for instance, Kerberos, require that a user authenticate to a trusted third party, such as a key distribution center. When a user's identity is verified, he is granted access, or a "ticket," to systems and resources for which he has authorization from ticket granting servers for a fixed amount of time [10].

- Digital signatures and certificates (described in Chapter 2).

- Smart card and token systems.
- Biometrics.

Authentication is discussed in more detail in Chapter 8.

5.6   Authorization (Access Control)

Both individuals (humans interacting with a computer) and applications running on systems can perform numerous functions on networks and computers. An individual can open a Telnet connection to a router, and execute commands from a command line interface over that connection. The same tasks the user performs can be scheduled and scripted using PERL, C++, or similar programming languages.

An authorization is a policy that specifies whether an identity has permission to access a resource (asset). Authentication provides identity verification, and governs the actions user or system is allowed to take, based on the trust asserted following authentication.

An access control is a means of implementing and enforcing authorization policies. Through an access control, we grant a user permission to perform, or prohibit the user from performing, an action on a network, computer, application or data object, as dictated by a security policy.

Access controls can be implemented by associating permissions to different kinds of identities, using names and addresses at every level of the Internet architecture. For example, access controls can be applied to Ethernet MAC addresses to grant or deny permission to forward packets through Ethernet switches and WLAN access points. Access controls at firewalls can permit or deny traffic flows between two or more networks based on source and destination IP addresses and Internet port numbers. Access controls at application proxies can be applied using user and group account names from an active directory, or based on URIs and URLs in SIP and HTTP messages. Operating systems access controls (Windows (NTFS) and UNIX file systems) can specify whether a user or member of a group account may read from and write to specific directories, folders and files, and execute application programs and binaries. Application level access controls may also govern which voice services a user is entitled to use.

5.7   Auditing

Auditing is the process of testing and verifying that the security measures an organization implements enforce the security policy defined by the organization.

The auditing process subjects individual host and overall network security design to careful examination. During an audit, an organization or trusted auditor tests every element in a network to verify that the measures taken to eliminate threats to networked assets are present and functioning as the organization intends.

Effective and thorough auditing is not merely a paper exercise. Analysis of a security design, architecture, and comprehensive configuration analysis represent a small part of the auditing activities. These are often complemented with an active vulnerability assessment, or penetration test, a process where assets are subjected to attacks, and flaws or problems in the security design and implementation are revealed. Auditors and penetration testers automate part of this process by using assessment tools that attempt to exploit known or common vulnerabilities of the operating and file systems used on the client and server systems, routers, switches, access points and security systems specific to your network. Skilled auditors will attempt to break into your network with only the knowledge obtained using the same methods as motivated attackers. Such zero knowledge attacks often reveal information your organization makes public that can be used to attack your networks. Auditors may also begin with partial (for example, access to a single user or administrative account) or even full knowledge so that they can emulate attacks that a motivated attacker might be able to perpetrate if assisted by an insider (an employee who has been manipulated, or coerced, or is disgruntled or recently dismissed).

Audits adhere to a formal test methodology. Each test is fully documented so it can be repeated exactly as it had been previously performed. Test documentation should include the breadth and scope of tests to be conducted, the expected versus actual results, analysis of results, and recommendations for remedying incorrect or undesirable results.

The purposes of conducting an audit are to confirm your assets are as secure as your policy intends, to identify where you are still vulnerable and, in some situations, to obtain an independent verification of your security (the latter may be required by potential e-business partners, insurance underwriters, or regulatory agencies). Independent auditors are often good choices for performing an audit. They have fewer biases and vested interests, and presumably less insider knowledge about your business processes and politics.

Audits often reveal problems, and organizations must be prepared to correct policy, process, design, configuration, and implementation based on the audit results. Further testing may be required to determine if the problems identified have been resolved, and if new vulnerabilities have been introduced during the resolution process. (The need to retest underscores the importance of documenting and designing repeatable test procedures.) Auditing may seem like an endless task, and in fact, it should. Security is an ongoing process and requires constant evaluation, testing, and reevaluation [6, 7, 11].

5.8   Monitoring and Logging

Implementing security isn't something an organization gears up for and performs once, and then sets aside with absolute confidence that its assets are protected and online presence is assured. System, application and network level events should be constantly and routinely monitored to determine if the elements of the organization's network continue to operate in the manner expected, within the anticipated ranges of performance. Real-time monitoring may reveal unauthorized activities, unapproved services operating on servers, suspicious processes running on client devices, or DoS attacks emanating from or the organization's internal subnets and probing its servers. Other monitoring may reveal configuration errors, as for example, traffic passed in the clear that should be protected using an encrypted tunnel. Still other monitoring may alert you to the fact that tampering has occurred with some critical information (web pages, configuration data, password files).

Some monitoring can be performed by "eyeballing" graphical displays of network traffic processed by switching and security systems, and CPU and memory usage, running processes, and applications on servers. Other monitoring is alarm-driven: detection of a suspicious or unauthorized activity causes an audible or visual alarm, or a page, to be issued.

Many forms of event monitoring can be automated. Security systems, including LAN traffic analyzers, intrusion detection systems, and firewalls can record events in real time and can be configured to generate alarms, pages, or email based on events you determine to be worrisome or that merit administrative intervention. Security software can be configured to alert administrators to tampering attempts against file systems, configuration abuse, and toll fraud.

Organizations typically incorporate a number of monitoring techniques into their daily security operations. However, maintaining sufficient vigilance through monitors so that administrators are notified of significant events, and are able to deal in real time with a potential onslaught of alarms, pages, and alerts, is always a balancing act. When defining a monitoring process, it's important to evaluate and prioritize events based on the significance of the asset and the threat an attack poses.

Logging is a process of continuously recording system and application level events, and network traffic. Log information collected on a network is analogous to the black box flight recordings on commercial jets, or the recorded output of a cardiac monitor. Logs provide a chronology of how a network performs and how it is used, or misused, over time. Accumulating logs over time provides a histogram of network activity. Organizations can use this record of events as a baseline, to distinguish normal activity from activity that is new, suspicious, or abnormal.

Log analysis reveals what traffic and events are normal for your network, based on the policy you intend to enforce, and the configuration and security measures you have implemented. Security staff that routinely review logs are likely to note trends, deviations from norms, and through such analysis, they may be able to anticipate impending security incidents and take measures to avoid them [12].

Applications, operating systems, security and switching systems all provide some event auditing and logging capabilities. Generally speaking, you should log events that help you determine how well your security enforcement mechanisms are performing. For example, an online brokerage may require that every its online trading application timestamp and log every transaction, successful or failed; that its web and database servers timestamp and log application, service, and process activity, and perhaps file I/O operations; and that its security systems (firewalls and VPN gateways) log the disposition of all traffic they process.

Most operating systems, servers, and security hardware and software support some form of security, application, or network event logging by recording events to log files. While log entry formats vary among products, most contain sufficient information to identify the who (user), what (the datum accessed), when (time stamp), where (server) and why (read, modify, delete) of an event or transaction.

Most systems that perform logging will do so locally and remotely; by forwarding log entries to a server using a proprietary logging protocol, or the syslog service [13]. The danger of only logging locally is that one of the first acts a skilled attacker will perform if he gains administrative control of a system, is to erase logs or modify them to remove traces of his activities. The danger of only logging remotely is that you may lose log data should connections between your logging devices and log server be severed. You may also need to use a VPN to secure the communications that transfer logging information.

Logs should be archived frequently as they can play a vital role in any criminal or civil suits you may seek against intruders. You must be able to prove that logs are unaltered and represent an exact record of a chronology of events, and courts will require that you show a chain of custody that cannot be repudiated during a trial [14, 15].

5.9   Policy Enforcement: Perimeter Security

Perimeter security is commonly implemented to separate a privately operated IP network from the public Internet. This divides access and privileges into policy domains. The demarcation point(s) are often collocated at the access point(s) connection to the Internet, where routers, switches, and firewalls are used to enforce policy. The traditional notion of a secure perimeter, however, has evolved considerably in recent years. The roles of security systems in enforcing policy, especially the roles of firewall systems, have diversified and become specialized. These roles are given some consideration in this book as they are prominent in providing security for voice and data networks.

5.9.1   Firewalls

The firewall is the most commonly used security system, and can take many forms. Standalone hardware security appliances, security software that operates on a general-purpose operating system (various flavors of Windows and UNIX), firewall-enabled routers and switches can be deployed individually or in combinations to separate and protect an internal network and trusted community from an external network and a community that is not unilaterally trusted.

A firewall enforces authorization policies on traffic between policy domains. In a security policy, the following should be defined:

- What types of traffic are permitted;
- The users who are allowed to pass traffic through the firewall;
- The systems from which traffic is allowed to pass through the firewall;
- Any other restrictions that are imposed, such as type of content, time of day, authentication required.

A firewall compares inbound and outbound traffic that passes through it against this set of rules. A firewall is also responsible for auditing, or keeping track of how the security policy is being enforced. Firewalls audit by recording security-related events in logs. Blocked traffic, as well as allowed traffic, can be logged, along with information that identifies the type of traffic, traffic origin/ destination, timestamp, and disposition (permit or deny).

Firewall rule sets are declarative sentences enforced by the firewall. As indicated in Figure 5.1, the security policy states that we don't want telnet connections, so we deny outbound traffic from trusted networks A to B to TCP port 23 on any host of any external network. Our rules also state that every system needs the DNS service, so we allow outbound traffic to any external name server from A and B. Finally, we want visitors to access a public-facing web site, so we allow http/80 traffic inbound from any external network, but only to the address of our public web server.

In addition to policy enforcement and auditing, many firewalls perform these security-related functions:

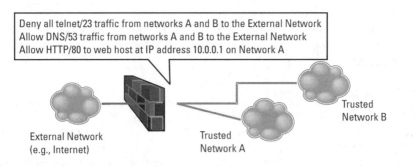

Deny all telnet/23 traffic from networks A and B to the External Network
Allow DNS/53 traffic from networks A and B to the External Network
Allow HTTP/80 to web host at IP address 10.0.0.1 on Network A

External Network
(e.g., Internet)

Trusted
Network A

Trusted
Network B

Figure 5.1 How firewalls enforce policy.

- Network address and port translation (NAT/PAT, see Section 5.10);
- User authentication (see Chapter 8);
- Protocol anomaly and intrusion detection;
- DoS prevention;
- Application content filtering (e.g., web URL filtering);
- Content blocking based on S/MIME type;
- Virtual private networking capabilities;
- Adaptive routing.

Several types of firewall implementations are currently in use.

5.9.1.1 Packet Filter

The simplest implementation of a firewall is a basic packet filter. A packet filter examines the IP, TCP, and application headers of each packet individually, and allows or blocks traffic based on the rule that matches this single packet. Static packet filtering checks individual packets for correct header composition, and applies filtering rules against a 5-tuple of packet header information consisting of:

- Source IP address;
- Source port number;
- Destination IP address;
- Destination port number;
- Higher-level protocol as specified in the IP PROTOcol field, such as TCP, UDP.

Because no state information is maintained across packets that comprise a flow, as for example, TCP session this firewall implementation is called a static packet filter.

5.9.1.2   Dynamic Packet Filter

Dynamic packet filtering allows a firewall to apply more complex policies by keeping track of flows and sessions established through the firewall. The firewall keeps track of sessions and flows that pass through it. How a firewall reacts to any packet it processes is influenced by other packets it has already seen, associated with this session or flow. This dynamic processing requires that the firewall maintain state for traffic flows, so dynamic packet filtering is also called stateful filtering. One benefit of dynamic packet filtering is that this kind of firewall can block attacks that span more than one packet. For example, a dynamic packet filtering firewall can detect and block attacks that attempt to hijack a TCP session, certain denials of service attacks, and so forth. Dynamic packet filtering has evolved over time to provide application level traffic inspection. Many forms of application protection, including content blocking based on S/MIME type, web application attack signatures, antispam detection, and gateway antivirus and antispyware capabilities are now performed by stateful firewalls.

Figure 5.2 provides an example of dynamic packet filtering. The relevant firewall rule set maintained by the firewall for this example are: to deny outbound FTP sessions initiated by all hosts on network A, and to allow inbound FTP sessions to the public FTP server at 10.0.0.10 from any external host.

In Figure 5.2, Dave wants to get a file from a public FTP server on network A. Dave's PC opens an FTP session to the public FTP server. The firewall allows FTP inbound, according to the rule set, and remembers this action by saving the FTP PORT command sent to the public FTP server. Dave requests a file (FTP GET) from the public FTP server. The public FTP server opens an FTP DATA session to Dave's PC. The firewall allows an outgoing session request to Dave's PC at the named port because it is stateful so it saved information from the previous FTP PORT command. A static packet filtering firewall

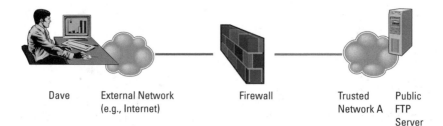

Dave   External Network   Firewall   Trusted   Public
          (e.g., Internet)   Network A   FTP
                                                                   Server

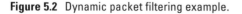

Figure 5.2 Dynamic packet filtering example.

would not have this state information, and would have enforced the default rule denying outbound FTP connections.

5.9.1.3 Application Proxy Firewall

An application proxy firewall reassembles all TCP or UDP segments of an application message, whether a DNS query, response or zone transfer, an e-mail message, an HTTP or SIP request or response, and examines the message in its entirety before relaying the message to its intended destination. Proxies are securely coded versions of filter and block application services. Secure proxies provide protocol anomaly detection, detect attempted misuse of the services an application provides, and many can filter and block application traffic based on message content, type and operation. For example, a secure SMTP proxy typically blocks attacks that use exploitable and malformed commands. An SMTP proxy may "scrub" and alter outgoing mail headers to generalize mail addresses to public domain names such as "corecom.com" rather than expose an internal mail server such as "mail1.corecom.com." It may also strip message attachments based on banned MIME types, virus and spyware signatures and spam patterns. Similarly, a secure HTTP proxy may block HTTP requests based on URL or MIME type and block incoming responses that contain content types that violate an acceptable use policy. A secure DNS proxy can restrict operation and query types (e.g., allow or deny a full or incremental zone transfer). As the market for secure SIP proxy servers matures, they will undoubtedly perform similar security services.

Figure 5.3 illustrates how an application proxy firewall might process HTTP request and response messages.

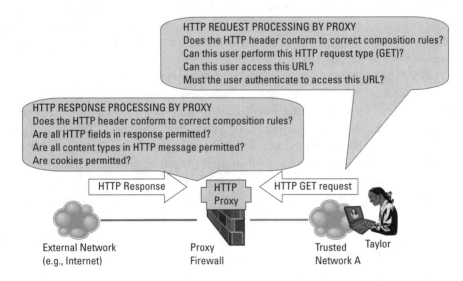

Figure 5.3 Application proxy firewall.

5.9.1.4   Common Firewall Implementations and Deployments

Some firewall systems support stateful inspection and application proxies. Some firewall deployments use multiple systems, routers and firewall appliances. A security administrator may perform very general packet filtering at a router and finer, stateful inspection and proxying at a firewall appliance. This configuration, called a screen and choke, is used to create a demilitarized zone (DMZ) where public-facing servers can be separated from trusted networks yet still be protected according to a security policy that accommodates e-commerce and public web services. Many firewalls provide additional LAN or virtual LAN (VLAN) interfaces for DMZs. Here, the screen policy is applied to traffic forwarded between the public facing (external) and DMZ networks, and the more stringent choke policy is applied to traffic forwarded between the public-facing and internal (trusted) networks. See [16, 17] for classic works on firewalls.

The primary objectives of an Internet firewall is to protect the organization's internal networks and information assets from unauthorized access by outsiders and to restrict authorized outsider (customer, business partner, guest) access to specific hosts and services. Another important objective of Internet firewalls is to control which Internet sites and services that insiders (company employees) may visit, and to control the types of content insiders may bring into the internal network.

A less common, but important, use of firewalls is to enforce a security policy between departments or business units as a means of isolating and segmenting networks according to security policies. In very large organizations, firewalls may be placed between a "core" organization and its acquisitions, divestitures and joint ventures. The primary reason to use firewalls in this manner is to isolate, or compartmentalize, groups and the sensitive data they handle from everyone else in the organization. This reduces insider threats by moving access controls closer to sensitive data.

Firewalls are very practical solutions in situations where it is important to separate wireless LANs from trusted networks, especially when supporting user authentication. The authentication methods that can typically be enforced through, or facilitated by, a firewall are typically stronger than those enforced by WLAN protocols alone. When such firewalls are used in combination with virtual private networking, the encrypted tunnels supported between wireless users (or generally, any user communicating from a server protected by that firewall) is an effective deterrent against attacks that involve traffic capture, replay, and analysis (sniffing).

5.9.2   Session Border Controller

Session border controller (SBC) is a loosely defined term for devices that combine a VoIP application layer gateway with media relay and firewall functions. SBCs can be used to filter and control VoIP sessions at the "border" between a public and private VoIP network (or two private VoIP networks). A SIP-based VoIP application layer gateway utilizes a back-to-back user agent (B2BUA). In a public/private scenario, a B2BUA acts as a user agent server on the public side and responds to SIP requests originating from client UAs via the public network. The B2BUA applies policy, and then relays the request to a server on the private VoIP network (as if it were a client UA originating a call on the private network). A B2BUA uses different `Call-IDs`, `To` tags, and `From` tags on the two halves of the session and maintains a map between the two (see NAT, Section 5.10). A B2BUA may also relay media sessions.

A B2BUA hides the internal topology and IP addressing from VoIP devices on the external (in our example, public) network. Internal IP addresses that are present in the `Via`, `Contact`, and SDP are rewritten, and additional `Via` header fields indicating the location of internal proxy servers are removed. This behavior is similar to network address translation, described in Section 5.10.

Functions commonly associated with session border controllers are:

- Topology and IP address hiding;
- Call admission and bandwidth control;
- Traffic management;
- Media relay and transcoding.

Session border controllers offer several security advantages, the most obvious being that voice signaling and media session processing can be filtered according to a security policy. The use of a VoIP application layer gateway can introduce some issues, depending on the topology and degree of sophistication (feature richness) of the gateway, including:

- *Single point of failure.* VoIP networks use a very scalable distributed architecture, but if all signaling and media traffic are routed through one a single SBC, that system can become the focus point for attacks. This can have a more severe impact than an attack on a single proxy server, which affects signaling only, or on a particular VoIP endpoint, which impacts that endpoint alone. The introduction of high availability and load balancing can diminish this threat.

- *Disruption of end-to-end security measures.* Many end-to-end security measures in a protocol such as SIP will be broken by the presence of an ALG. For example, an `Identity` signature will not validate after the request has been routed through an ALG as the dialog information will have been changed. The end-to-end architecture of the Internet is extremely important [18] and should not be broken without extreme care.

- *Reduced feature set.* Since the ALG is an endpoint in the session, the ALG must understand every feature and service used. If two endpoints both support a feature but the ALG does not, the endpoints cannot make use of this feature. While it is often useful to support a minimum feature set in application proxy firewalls, in this situation, a reduced feature set may be overly constraining.

- *Reduced service quality.* The introduction of a relay point in a signaling session is unlikely to add noticeable latency, but unless service quality is accommodated by the ALG, it may introduce meaningful latency in media sessions.

5.9.3   Firewalls and VoIP

VoIP poses three challenges to firewall deployment:

1. VoIP uses dynamic ports for media sessions. Firewall administrators are reluctant to open hundreds of ports for real time traffic, but traditional firewalls aren't very good at opening and closing ports dynamically

2. Firewalls provide network address translation, but address modification interferes with VoIP signaling.

3. Many firewalls are not QoS-aware. Some aren't designed to process the traffic patterns exhibited by VoIP applications, such as large numbers of small packets requiring predictable latency and jitter characteristics. Toll-quality voice requires a firewall solution that does not introduce noticeable latency and jitter. This is especially difficult when encryption processing associated with security protocols is not performed in hardware.

A VoIP-aware firewall is an application-aware firewall that can process VoIP protocols, provide protection against VoIP-specific attacks, and can accommodate performance requirements for near-toll quality voice. Traditional network firewall security measures and even other application protection measures may be provided in the same firewall system.

Some VoIP-aware firewall solutions decompose the VoIP security problem into three elements. Decomposing the problem in this manner doesn't imply that three physical firewalls are required. Rather, the firewall logic handles VoIP security by solving the three problems aforementioned problems separately:

One component handles call signaling, authentication, access controls, and NAT. This can be handled well in a proxy because call signaling isn't as sensitive to QoS as the actual voice traffic transmitted during the call. This component may, or may not, include cryptographic protection of signaling.

A second component micromanages port assignment and dynamic port authorization. Conceptually, this port control mechanism opens only as many ports from the assigned UDP needed support media sessions.

The final component handles voice traffic. Figure 5.3 illustrates the components of this conceptual VoIP-aware firewall.

TLS is used in this example, but IPSec or other proprietary or standards-based security protocols could be used (see Chapter 6). While not depicted in Figure 5.3, all components perform application-specific attack detection and blocking.

An initial list of firewall selection criteria emerges from this conceptual model:

- A security protocol, to provide support for signaling and media streams with endpoint authentication, message and media encryption, and integrity protection;

- SIP-aware NAT and media port management;

- Granular call admission control (CAC);

- Ability to control the number of concurrent calls (DoS protection);

- Protect against IPT-specific application-level attacks;

- Ability to monitor for unusual calling patterns (IDS, IPS);

- Provide detailed logging of all VoIP-related events;

- Monitor and maintain QoS on all media packets;

- Provide priority handling to media packets and preserve QoS marking on traffic that flows through the firewall.

5.10 Network Address Translation

Network address translation (NAT) was originally designed as one of several solutions for organizations that could not obtain enough registered IP network numbers from Internet address registrars for their organization's growing population of hosts and networks.

Several forms of IP address and port translation are currently being used in IP networks.

Network address port translation (NAPT), also called dynamic NAT, IP masquerading, or simply port address translation (PAT) dynamically maps all the private addresses used on a network onto a single public IP address. This form of NAT is commonly used by firewalls to hide the trusted network IP addresses from the Internet. In this deployment, NAPT works as follows:

- A host on the network protected by a firewall sends a packet out with a private source IP address [19] (e.g., 10.0.0.1, 172.16.24.3, 192.168.0.14). This packet also carries a source port number (e.g., 3506).

- When the firewall receives the packet, it changes the private source IP address to the IP address assigned to the public-facing network interface of the firewall (206.254.208.100).

- The firewall also changes the source port number (3506) to an unused source port number on the firewall (6000).

- The firewall maintains a mapping between these port numbers in its memory so it can process response packets.

When a response packet destined for IP address 206.254.208.100, destination port number 6000 is received by the firewall, the firewall maps this information back to the private source IP and port number that issued the original packet in this TCP connection or UDP stream.

Only one public address is revealed to the world outside, but this single address represents every host on your network. How economical. Hundreds, even thousands of hosts can share the same public address.

Static NAT not only has several names, but several variants as well. One-to-one (1:1) static NAT can be used at an Internet firewall, as for example, to bind a unique public address to each privately addressed host protected by that firewall. With this binding in place, the firewall forwards traffic between the address assigned to a server from your trusted or DMZ network address space, and a public address that can be obtained via a DNS query anywhere in the Internet. Consider the following example. You assign the private address 10.0.1.11 to a server on a trusted or DMZ network protected by a firewall, and then bind a public address 206.254.208.100 to this private address using 1:1 NAT at your firewall. You create a DNS entry in a publicly accessible DNS server that maps www.buykewlstuff.com to 206.254.208.100. All DNS queries for www.buykewlstuff.com resolve to 206.254.208.100. All packets with this destination address are delivered to your firewall, which forward them to the private address 10.0.1.11. Using static NAT in this manner, you can grow your servers to a whole farm (www.buymorekewlstuff.com, www.buyevenmore kewlstuff.com), assigning each of them unique private and public addresses and binding them in this fashion at the firewall.

1:1 static NAT forwards one or all ports to a server behind your firewall. Port forwarding is another form of static NAT that is used to selectively map individual ports to individual servers. Suppose you wanted to allow HTTP access to a private server at 10.10.11.11, but block all other services. Using port forwarding, port 80/HTTP traffic delivered to the firewall's public address will be forwarded to 10.10.11.11, port 80 (or to any port you choose). Unlike dynamic NAT, static NAT provides no addressing economy. However, both forms of static NAT are useful when you want to protect your public-accessible servers or specific services from attacks firewalls are generally designed to defeat [20].

5.11 Intrusion Detection and Prevention

Intrusion detection systems (IDSs) are useful and effective additions to Internet security defenses. As the name suggests, they are a class of software and hardware that is used to alert network operators to an active or impending attack against some networked resource.

There are two types of IDS: scanners and monitors. Scanners are also called vulnerability assessment and security benchmarking tools. Scanners look for known security problems on a host that make it vulnerable to attack, namely, default accounts, dangerous file permission settings, lax security policy settings, missing security patches, and configuration choices. Scanners can also check to see whether important files have been removed or modified. They can alert administrators of vandalism, for example, whether malicious code (root kits, remote administration tools) has been installed on systems. Scanners are available for most major client and server operating systems, and many can also scan for vulnerabilities in commonly used server applications.

Scanners can test and report on individual hosts or any (all) hosts on a network. Network scanners identify the network services offered by individual computers on a network, and report what kinds of information these services reveal about the server, service, user account, and file system configuration. Web scanners identify vulnerabilities specific to web servers: default accounts, default and exploitable CGIs, vulnerabilities to cross-site scripting [21, 22], and other URL-based attacks. System integrity scanners make a cryptographic snapshot of a system. This is used to determine if important file system configuration files, system programs, or even web pages have been modified. Scanners are proactive assessment tools that help administrators eliminate vulnerabilities before they can be exploited by attackers, and the lax security and configuration they might reveal to them can cause serious problems.

IDS monitors are dynamic analysis systems. They capture and examine network traffic in real time and generate alarms if captured traffic patterns suggest or clearly identify that a network attack is in progress. They also maintain detailed logging information of all network traffic that can be used for trending, predictive and post-attack (forensic) analysis.

Two kinds of IDS monitoring and real-time analysis methods are currently being deployed: misuse detection and anomaly detection.

There are two types of misuse detection systems. For the first type, an administrator identifies a list of things that should not happen, and then the IDS monitor watches for these events. "What should not happen" is based directly on the network security policy. For example, if the security policy says only HTTP, FTP, and SMTP are permitted from the Internet through the firewall, a misuse system watches for other types of packets from the firewall. This is difficult for an attacker to deceive. The second type of misuse detection system is also called an attack signature recognition system. Misuse or attack signatures are first codified, and then a data source—a network telemetry system or an operating system audit log—is monitored for patterns of attack. A user-level process that starts up and acquires system or "root" privileges without executing the "su" (set user) command is an example of a simple misuse signature on a Unix system.

An anomaly detection system is told, or actively "learns," what is considered normal behavior for an individual, a system, or a network, and takes action when some event falls outside of the normal range. Basically, they let us know when "something is fishy." While people can be trained to do anomaly detection very well, it's very difficult to do by computer.

Much research has been done in the area of anomaly detection, but only very simple anomaly detection systems are in real use today. Disk usage growth or shrinkage outside of a certain rate per minute can be tagged as an anomaly. Individual user activities outside of normal use hours, or connections to the network that are not from the user's usual machines are easily flagged as anomalous behavior. Sophisticated systems, where, for example, an individual user's typing patterns or network use patterns are "learned" are not here yet.

Some products incorporate both analysis methods. Intrusion detection systems are widely used today. Some of the current generation of network IDSs work in concert with other security systems; some, for example can direct firewalls to modify security policy and block sources of suspicious or malicious traffic. Early IDS systems relied on passive probes and were primarily used for detection and notification. Newer IDS systems can run "in line" and examine traffic at gigabit speeds. Some of these in-line IDS systems can detect and block attacks and describe themselves as intrusion prevention systems. Some focus on blocking distributed denial of service attacks [23, 24]. IDSs are a constantly evolving security component that attracts considerable attention and research. They are too often relied upon as a substitute for expertise and diligent application of other proactive security measures. Just as it is imprudent to rely entirely on firewalls to protect networks and take no measures to harden servers, it's just as foolhardy to rely entirely on IDS to detect and block attacks. IDSs are good tools, but ultimately, they are but one component of security and play one role in a security system.

5.12 Honeypots and Honeynets

A "honeypot"[25] is a passive security measure that uses deception techniques to observe and record the activities of an attacker. Honeypot programs can assume the appearance of an application service or a host. While it may appear to be exploitable, a honeypot is really a well-contained environment investigators use to lure and entrap an attacker, and then contain his activities. The objective of a honeypot is to observe an attacker without detection long enough for him to reveal what he's after, and while accumulating evidence sufficient to track him down and prosecute him in a criminal or civil court.

A "honeynet" [26] is an extension of a honeypot. Security and law enforcement investigators will set up an entire network to create an even more elaborate deception than a honeypot. Honeypots and honeynets complement intrusion detection systems. Honeypots and honeynets record every action taken by an attacker. They log access attempts, capture keystrokes, identify files accessed and modified, and note programs executed. In short, they spy on, and record, the attacker's every activity.

The most critical aspect of successful honeypot deployment is deception. A honeypot must look exactly like the service or resource under attack. A password file, service banner, and file permissions must look correct, but improperly configured against attacks. The deception must be sustained for however long the investigators hope to observe an attacker. This can become quite involved. Investigators may periodically add and remove files that seem to hold sensitive information, perform routine (but sloppy) maintenance, and so on.

One of the most practical reasons to operate a honeypot, especially in a VoIP network, may be to identify stolen accounts. Your honeypot can help identify any accounts an intruder has attempted to use, as well as those he's succeeded in breaking into. Since VoIP equipment and applications are relatively new, another important application for honeypots in VoIP networks is to discover attack techniques and expose vulnerabilities on a honeypot system and take measures to mitigate the vulnerability on production systems before it is widely exploited.

Honeypots require considerable security expertise and time. Improperly configured or deployed honeypots could expose one of your production hosts. Weigh these concerns carefully before you play cat-and-mouse with attackers.

5.13 Conclusions

The information provided in this chapter provides sufficient background material about security processes, practices, and technologies to allow the reader to learn about the measures and strategies applied to securing voice and data networks currently employed. Volumes have, and continue to be, written about security. The references provided in this chapter should allow readers at varying interest levels to complement what they have learned here in order to achieve their desired comfort level.

References

[1]   "Castles in England and Wales," http://www.britainexpress.com/History/castles.htm.

[2]   SANS Security Policy Project, http://www.sans.org/newlook/resources/policies/policies.htm.

[3]   ISO 17799 Directory: Services & Software for ISO 17799 Audit and Compliance & Security Risk Analysis, http://www.iso-17799.com/.

[4]   Network Security Library: Security Policy, http://secinf.net/ipolicye.html.

[5]   Virginia Department of Education, *Acceptable Use Policies: A Handbook*, http://www.pen.k12.va.us/go/VDOE/Technology/AUP/home.shmtl.

[6]   Information Systems Audit and Control Association & Foundation http://www.isaca.org/.

[7]   Wadlow, T., *The Process of Network Security: Designing and Managing a Safe Network*, Reading, MA: Addison-Wesley, 2000.

[8]   Piscitello, D., "Security Hats: Black or White, No Grayscale," http://hhi.corecom.com/blackorwhitehat.htm.

[9]   BBC News, "Passwords Revealed for Sweet Deal," http://news.bbc.co.uk/1/hi/technology/3639679.stm.

[10]   Neuman, C., T. Yu, S. Hartman, and K. Raeburn, "The Kerberos Network Authentication Service (V5)," RFC 4120, July 2005.

[11]   Amoroso, E., *Fundamentals of Computer Security Technology*, Englewood Cliffs, NJ: Prentice-Hall, 1994.

[12]   Stutzman, J., "Primer on Predictive Analysis," *TISC Insight*, http://www.tisc-insight.com/newsletters/313.html.

[13]   Lonvick, C., "The BSD Syslog Protocol," RFC 3164, IETF, http://www.ietf.org/rfc/rfc3164.txt.

[14]   Matsuura, J., "Managing Electronic Evidence," *TISC Insight*, http://www.tisc-insight.com/newsletters/321.html.

[15]   Matsuura, J., *Security, Rights, and Liabilities in E-Commerce*, Norwood, MA: Artech House Publishers, 2001.

[16]   Bellovin, S., and B. Cheswick, *Firewalls and Internet Security*, 2nd ed., Reading, MA: Addison-Wesley, 2004.

[17]   Zwicky, E., S. Cooper, and D. Chapman, *Building Internet Firewalls,* 2nd ed., Sebastopol, CA: O'Reilly & Sons, 2000.

[18]   Carpenter, B., "Architectural Principles of the Internet," RFC 1958, June 1996.

[19]   Rehkter, Y., D. Karrenberg, G. J. de Groot, and B. Moskowitz, RFC 1918, "Address Allocation for Private Networks," RFC 1918, IETF, http://www.ietf.org/rfc/rfc1918.txt.

[20]   Piscitello, D., "How and When to Use 1:1 NAT," *Core Competence*, http://www. corecom.com/external/livesecurity/1to1nat.htm.

[21]   Piscitello, D., "Anatomy of a Cross Site Scripting Attack," *Core Competence*, http://www. corecom.com/external/livesecurity/xscript.htm.

[22]   The Cross Site Scripting FAQ Page, http://www.cgisecurity.com/articles/xss-faq.shtml.

[23]   Dittrich, D., "Distributed Denial of Service Attacks/Tools," http://staff.washington.edu/ dittrich/misc/ddos/.

[24]   Farrow, R., "Distributed Denial of Service Attacks," *TISC Insight*, http://www.tisc2002. com/newsletters/216.html

[25]   Spitzner, L., *Honeypots: Tracking Hackers*, Reading, MA: Addison Wesley, Pearson Education, Inc., 2003.

[26]   Piscitello, D., "Honeypots: Sweet Idea, Sticky Business," http://www.corecom.com/external/livesecurity/honeypots.html.

6

Security Protocols

6.1   Introduction

To continue with the castle analogy from the previous chapter, it is clear that the stationary defenses of a castle proved very effective so long as the treasures weren't moved and the population of the kingdom didn't venture beyond the security the castle provided. For nobles and merchants, however, travel, trading goods, and communication with other kingdoms were economic and political imperatives. Security had to be provided for these personages and their goods, while they were in transit. Many measures were taken to protect the noble's entourages and the merchant's trade wagons. They were accompanied by armed guards and knights and their treasures were transported in strongboxes. Private correspondence was stored in protective cases, sealed, and uniquely imprinted with wax and chop (or signet ring).

In much the same way, IP networks, and hence their security, were also initially based on isolationist practices. Much as isolated kingdoms proved impractical, so are wholly isolated, private networks today. Most organizations must have an Internet presence, and their employees must access Internet resources, from the office, at home, and while traveling. In addition, many organizations have mobile workforces. Thus, every organization has information assets that must be protected from misuse, abuse, theft, or damage from outsiders. Increasingly, organizations allow business partners, customers and consumers to access information via intranets and extranets. Organizations exchange sensitive correspondence and perform business transactions electronically, over the Internet, as well. These organizations are growing more aware of the threats Internet-originated attacks pose, and want to protect access to their information assets, and to protect information exchange over the Internet of as well. To accomplish this, additional security measures are often required.

One of the most widely employed security measures is virtual private networking. A virtual private network (VPN) combines two networking concepts: virtual networking, and private (secure) networking [1]. Virtual networking provides a service to a community within a given addressing and routing domain. It allows geographically distributed users and hosts to interact and be managed as a single network. Virtual networking extends the user dynamics of LAN or workgroup beyond a single physical location and independent of physical connectivity and topology. A virtual LAN (VLAN) is an excellent example of a virtual network. In a VLAN, stations connected to different physical LANs in different offices in different cities are addressed and traffic is routed as if all the stations were connected to the same physical LAN in the same office.

A VPN uses security protocols to protect information exchanged over the Internet, or generally, any communications path that is considered untrustworthy. Security protocols apply cryptographic measures (discussed in Chapter 2) to authenticate users and endpoints and to prevent information in motion, including voice conversations, from being eavesdropped upon, modified, and replayed.

Following are the most commonly used Internet security protocols:

- IPSec (Internet Protocol Security) provides authentication, encryption, and integrity services at the IP layer. When IPSec is used, IP traffic is routed or tunneled between endpoints that are separated by an arbitrary network topology. Tunneled packets are wrapped inside IETF-defined headers that provide endpoint authentication, message integrity and confidentiality. These IP Security (IPSec, [2–4]) protocol extensions, together with the Internet Key Exchange (IKE, [5]), can be used with many authentication and encryption algorithms.

- TLS (Transport Layer Security), also known as Secure Sockets Layer (SSL), provides the same set of security services as IPSec, but at a shim or middleware layer between Internet's transport and application layers for applications that utilize TCP (Transmission Control Protocol) transport. Servers are authenticated using digital certificates, and clients can be authenticated using a variety of authentication methods.

- Datagram TLS (DTLS) is a new protocol, an adaptation of TLS that can be used to secure User Datagram Protocol (UDP) exchanges.

- The Secure Shell (SecSH, SSH) also operates between Internet's transport and application levels. SSH was originally developed as a secure replacement for UNIX "r" commands (rsh, rcp, and rlogin), but is used for more than remote system administration today. Generally speaking, any application protocol can be tunneled over an authenticated and encrypted SSH connection.

All security protocols require certain information prior to creating tunnels for secure information transfer:

- The identities of the peers involved in the data transfer;
- The cryptographic keys associated with the identities;
- Resources to be protected by the employment of a secure tunnel for communication;
- The cryptographic algorithms to be used and how they will be used (confidentiality, integrity, non-repudiation);
- The keys associated with these algorithms.

Some of this information can't be sent unprotected or the security of any subsequent data transfer would be compromised. All security protocols address this problem by using special protocols to identify properties of secure data transfer between peers. As examples, IPSec has the *Internet Key Exchange* Protocol, TLS has a *Handshake* Protocol, and SSH has a *Transport Layer and User Authentication* Protocol. Communicating endpoints use these protocols to mutually agree on such things as hash algorithms, bulk encryption algorithms, authentication methods, and key lifetimes. They also provide endpoints with a means of securely exchanging initial keying material and identity information. Keys that will subsequently be used to protect and authenticate messages transferred across the secured tunnel are exchanged or derived without exposure to attack.

This chapter introduces security protocols, their management and key exchange protocols, and discusses how they can be used to secure VoIP signaling and media sessions. We also discuss two application-level security protocols, DNS security (DNSSEC) and pretty good privacy (PGP) because of their applicability to current and future secure VoIP deployments.

6.2 IP Security (IPSec)

IPSec provides the following security services:

- Message confidentiality. Sensitive data are encrypted using one of several negotiated cryptographic algorithms.
- Source authentication. IP packets are signed by the sender and verified by the recipient.

- Message integrity. Modification of a message is detected by verification of the signature.

- Antireplay protection. IPSec detects when packets arrive out of order, much like TCP; however, out-of-order arrival is treated as an attack.

- Access control. When using IPSec, you define filtering rules for traffic disposition (selectors). Packets may be protected using IPSec, dropped, or they may pass through without protection based on the composition of the filtering rule.

IPSec was developed both for IPv4 and IPv6. It has two main components, the Authentication Header (AH) and the Encapsulation Security Protocol (ESP). AH adds source authentication and message integrity to every IP packet. The Authentication Header is interposed between a packet's IP header and its payload. The AH header consists of an index, a sequence number, and keyed hash. The recipient uses the index to identify the hash algorithm and key used to sign the packet. The hash is computed over the entire IP packet, including header fields that do not normally change in transit. The packet recipient applies the same algorithm and key to reproduce the hash and verifies that it matches the value carried with the packet. Few IPSec deployments use AH because confidentiality is not provided.

ESP provides all the services AH provides and adds message confidentiality using symmetric cryptography. ESP can protect the payload of an IP packet, called transport mode, or the entire IP packet, called tunnel mode. When transport mode is used, a new IP header is prepended to the original packet header, and only the data portion of the original packet is encrypted. When tunnel mode is used, a new IP header is pre-pended to the entire original IP packet, and both header and data of the original IP packet are protected. Tunnel mode encrypts the addresses of the original IP packet header to hamper traffic analysis attacks. Only endpoints must process ESP in tunnel mode, whereas every hop in a routed path must process transport mode ESP. Like AH, the ESP header contains an index that the recipient uses to identify the packet sender, cryptographic algorithms and keys used to protect the encapsulated packet or payload. ESP encrypts the encapsulated packet or payload using symmetric key encryption. Like AH, ESP authenticates every IP packet, however ESP authentication des not protect the encapsulating IP header from modification.

Both AH and ESP can be used in transport and tunnel mode. Figure 6.1 illustrates tunnel mode and transport mode packet composition using ESP.

Transport mode is used for IPSec use between Internet hosts. Figure 6.2 depicts computers as endpoint devices, but VoIP endpoints could also use IPSec in this manner to secure calls between individuals.

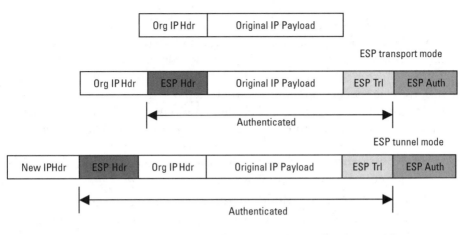

Figure 6.1 Comparison of tunnel and transport modes (ESP case).

Figure 6.2 IPSec transport mode.

IPSec is frequently used between two security gateways to provide security for entire subnets or subsets of hosts within the protected subnets. In this configuration security gateways commonly use tunnel mode, as shown in Figure 6.3.

6.2.1 Internet Key Exchange

IPSec endpoints use symmetric keys for several of its security services. To distribute these keys securely, and to periodically generate new session keys, IPSec endpoints use the Internet Key Exchange (IKE) protocol. IPSec endpoints use IKE

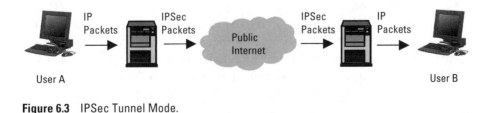

Figure 6.3 IPSec Tunnel Mode.

to authenticate peer security gateways as well. IPSec endpoint authentication uses pre-shared secret keys, public keys, or unsigned (raw) digital certificates (extensions to IKE provide support for other authentication methods as well [6]).

IKE performs these functions in two phases. In IKE Phase 1, peer endpoints exchange identities, authentication methods, and credentials to prove their identities. Following endpoint authentication, they negotiate security parameters to protect all future IKE communications from attack, specifying:

- Encryption algorithm is used to protect IKE messages from disclosure or capture (message confidentiality).

- Authentication algorithm (hashed message authentication code, HMAC) is used to verify the source of each IKE message and also to protect the packet against modification.

- Diffie-Hellman group is the method by which the local and remote security gateways will generate the same secret keys without revealing them. The peer security gateways use these keys in association with encryption and authentication algorithms they agree to use to protect SA traffic.

These parameters are collectively referred to as an IKE security policy. The secure channel defined by this policy is called an IKE security association.

A security association (SA) is a set of parameters that defines the security context for one-way communication between IPSec endpoints. For full duplex communications, two SAs are needed. Both IKE and IPSec use SAs to identify the security functions that must be applied between pairs of "tunnel" endpoints. The security parameters endpoints negotiate for individual IPSec tunnels include IPSec encryption and hash algorithms, IPSec Diffie-Hellman group (the basis for random number generation for keying material), protocol (ESP or AH), mode (tunnel or transport), and key lifetime/expiration for this SA.

IKE Phase I establishes a secure control channel. During a second or Phase II, IKE peers negotiate security parameters for individual IPSec security associations that will be used to secure data transfer. Hosts can multiplex many applications, including VoIP, over a single SA, or they can create unique SAs for individual applications and host-pairs.

The process of establishing IKE SAs can be performed in one of two modes. Main mode protects the endpoint identities from disclosure and is considered a more secure choice than the alternative aggressive mode, where fewer messages are exchanged but the identities are not encrypted. Aggressive mode is commonly used in remote access where the identity information is a dynamically assigned IP address. IKE operation is shown in Figure 6.4.

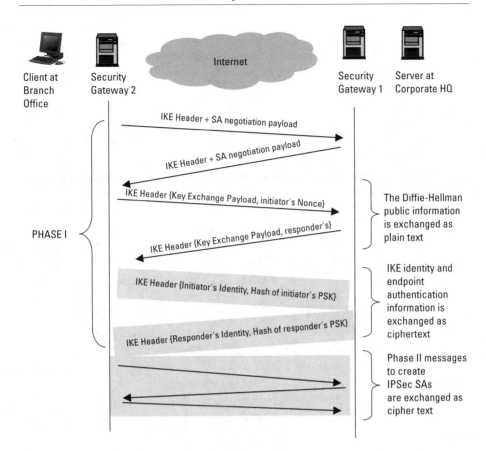

Figure 6.4 IKE main mode using preshared key authentication.

6.3 Transport Layer Security (TLS)

Transport Layer Security (TLS) [7] provides encryption, authentication, and confidentiality at a shim layer between the application and transport layer. Even more granular authorization as well as endpoint and end user authentication can be provided at the application level using TLS and other *secure stream* protocols such as SSH.

TLS is only used over TCP transport. TLS is the most commonly used security protocol on the Internet today.

TLS is based on Secure Sockets Layer (SSL) which was developed by Netscape Communications during the early days of the World Wide Web (WWW). The first version of TLS version 1.0 is based on SSL 3.0. The most recent version of TLS version 1.1 [8] incorporates some minor security fixes and clarifications.

The TLS Record protocol encapsulates higher level protocols and the signaling messages as well. TLS uses Handshake messages to establish security associations between client and server and define the cryptographic algorithm selection, initial key derivation material, compression selection; Alert messages to signal errors and session close; and ChangeCipherSpec messages to change the encryption set (CipherSpec) currently in use by TLS session peers. TLS applications use a Version field to make certain they operate the same level of the TLS specification. TLS computes an HMAC for message protection. TLS uses a Pad and Pad length field to support block ciphers. The Pad length tells the receiver how many bytes are Pad data and not actual application data. TLS supports the RC2, RC4, IDEA, DES, 3DES, and AES symmetric ciphers.

TLS Handshake is performed in two phases. In the first phase, endpoints agree on the encryption that will secure the connection. At this time, a certificate-based server authentication is always performed (mutual authentication using certificates is optional). TLS assigns an identifier to each session to support situations where the endpoints agree to suspend and later resume a session. During this phase, endpoints exchange a master key (keying material) that both parties will use to derive session keys for encryption and hash signing. Phase two of the SSL Handshake is optional. During this phase, the client is authenticated.

Client and server use the secure authenticated tunnel established during Handshake to protect application data they exchange.

The messages used in a TLS Handshake are shown in Table 6.1.

Figure 6.5 illustrates the flow of messages for a representative Handshake exchange.

At any time, the security characteristics can be changed by the client sending a ClientHello message and initiating a new TLS Handshake exchange. TLS continues to send and receive data using the current cipherspec until a ChangeCipherSpec message is sent and the cipherspec is updated.

A premaster secret is either exchanged or generated during TLS Handshake. When RSA mode is used, the premaster secret is generated by the client and sent encrypted to the server using the server's public key. When Diffie-Hellman mode is used, the DH exchange results in the generation of a pre-master secret. Both the client and server generate a master secret from the pre-master secret and the random strings contributed by the client and the server using the pseudo-random function defined in the TLS specification [7].

TLS allows a transport compression scheme to be negotiated during the TLS Handshake Protocol. The base specification only defines null (no) compression, but two extensions add the DEFLATE [9] and Lempel-Ziv-Stac (LZS) [10] compression methods.

TLS supports many endpoint authentication methods. Two endpoints can authenticate using mutual certificate-based authentication. Proxy servers might use this authentication method to secure SIP signaling (see Chapter 9).

Table 6.1

TLS Handshake Protocol Messages

Message	Description
HelloRequest	Can be sent by a server to ask the client to begin the TLS Handshake protocol. On opening a TLS connection, a client normally sends a ClientHello without waiting for a HelloRequest from the server.
ClientHello	First message sent by client to open connection. If used to reestablish an earlier connection, it contains the previous session ID. Also contains the version number of the protocol, timestamp and random sequence, list of supported cipher suites, and list of supported compression suites.
ServerHello	Server response to a ClientHello which also contains the protocol version number, timestamp, and random sequence, selected cipher suites and selected compression suites from the list supplied in the ClientHello message.
ServerCertficate	Sent by server after ServerHello. Contains the server's certificate, usually an X.509v3 certificate.
ServerKeyExchange	Sent by server after ServerHello if additional information besides the ServerCertificate is needed to exchange or generate a premaster secret. The format of the message depends on keying mode selected by the server in the ServerHello message.
CertificateRequest	Sent by the server after ServerHello to request the client provide a certificate for authentication.
ServerHelloDone	Sent by the server after the completion of ServerHello and other server messages which may immediately follow the ServerHello.
ClientCertificate	Sent by the client if the server requests a client certificate using the CertificateRequest message.
ClientKeyExchange	Sent by the client to complete the generation or exchange of the premaster secret.
Finished	Sent after a ChangeCipherSpec message which is encrypted using the new master secret.

Client-server applications, such as web and email portals, use an alternative form of mutual authentication sometimes called subauthentication. Conceptually, sub-authentication is subordinate to server authentication: the SSL/TLS server authenticates the SSL/TLS client *after* the client has authenticated the server. SSL/TLS subauthentication isn't confined to using digital certificates; applications may securely use a weaker authentication method such as a username and

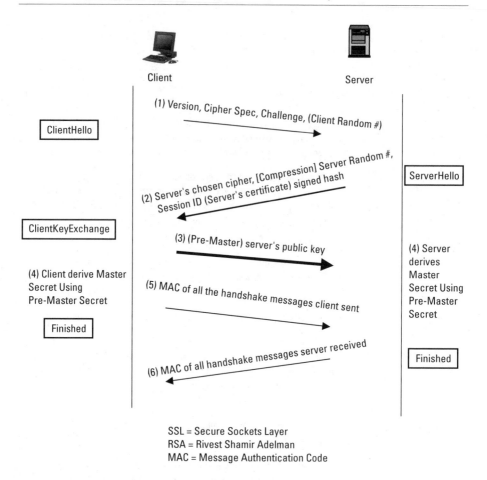

Figure 6.5   TLS Handshake Protocol.

password because challenges and credentials will be exchanged over the encrypted TLS tunnel. Subauthentication method support depends on the implementation and application, but every popular single and multifactor authentication method has been used with TLS.

Subauthentication is common in web applications in which personal, financial, or other important information is exchanged. The mutual authentication is performed using the steps below and illustrated in Figure 6.6.

- A TLS connection is opened from the client to the server. The server passes the certificate to the client during the TLS Handshake.

- The client validates the certificate by verifying, for example, that the DN (distinguished name) is the same as the web server's domain name, that the certificate has not expired, and that the certificate is signed by

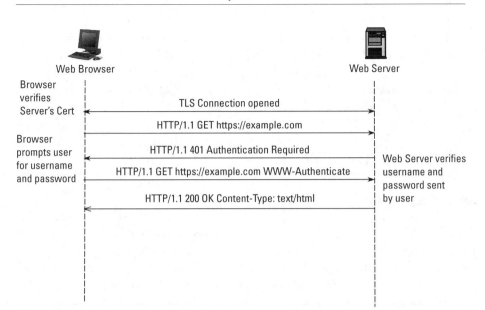

Figure 6.6   Subauthentication on the web with TLS.

one of the trusted CAs maintained by the browser. If the certificate passes this inspection, then the server is authenticated. (Further checks of certificate revocation lists, or validation using OCSP or SCVP can also be performed, depending on the application.)

- The server authenticates the client using a specified sub-authentication method, using a form on a web page which uses POST, or HTTP basic or digest authentication. The details of HTTP basic and digest authentication, which are also used to authenticate SIP signaling peers, are discussed in Chapter 9.

6.4   Datagram Transport Layer Security (DTLS)

Datagram TLS (DTLS) [11] is an adaptation of the TLS protocol that works over a datagram transport protocol, or User Datagram Protocol (UDP)[12]. DTLS attempts to reconcile two characteristics that make TLS unsuitable for running over UDP. First, the TLS Handshake assumes TCP provides a reliable transport and has no mechanisms to deal with packet loss and out-of-order delivery. Second, traffic encryption performed within the TLS Record protocol chains cryptographic context (block cipher state or stream cipher keys stream) across records. This means that if a record (datagram) is lost any records that follow cannot be deciphered. DTLS fixes both these problems by adding a retransmission scheme to handle datagram loss, adding explicit and independent state to each record, and by adding sequence numbers and a reordering function to the record protocol. DTLS also adds a fragment length and offset to accommodate the transmission of large Handshake messages (e.g., a ServerCertificate message). Some of these solutions are borrowed from IPSec ESP. An optional replay protection feature duplicates replay detection offered in IPSec AH and ESP.

At the time of publication, DTLS has been approved for publication as an RFC. It is potentially useful in securing both the SIP signaling and RTP media in a VoIP system (see Chapters 9 and 10).

6.5 Secure Shell (SecSH, SSH)

Secure Shell (SSH) was developed to provide secure remote login and a variety of services including secure file transfer. SSH is commonly used to replace these and provide a much higher level of security. Various implementations of SSH versions 1.x are available in commercial products and open source. The IETF has standardized SSH version 2.0.

The architecture and security analysis of SSH is described in [13]. The SSH protocol has three main parts and a number of extensions. They are SSH Transport Layer Protocol, SSH Authentication Protocol, and SSH Connection Protocol.

The SSH Transport Layer Protocol, SSH-TP [14] typically runs on top of TCP. A client uses SSH-TP to establish a secure connection to a server. SSH-TP provides server authentication, cryptographic algorithm negotiation, and key management. During the SSH-TP exchange, client and server negotiate key exchange method, public key algorithm (DSA or RSA), symmetric encryption algorithm for confidentiality services (3DES-CBC, AES-CBC), hash algorithm (MD5, SHA1) for integrity protection, and compression method (ZLIB, [18]). SSH supports a number of other encryption methods (Blowfish, Arc4, IDEA, CAST128) that utilize 128 bit or longer keys. SSH uses HMAC-SHA1 or HMAC-MD5 to provide integrity protection. The authentication key negotiated during the session is applied over a sequence number concatenated with the plaintext packet. The sequence number is an implicit 32 bit packet count that is initialized at 0 at the start of a SSH session then incremented for each packet sent.

Although version 2.0 supports key exchange method negotiation, most implementations use Diffie-Hellman key agreement to generate a unique shared secret key for each SSH session. SSH uses digital signatures over the key exchange messages for authentication but can also support shared secret key authentication mechanisms. The result of the DH operation is a shared secret key and an exchange hash. Client and server use the exchange hash as a session identifier for the duration of the session, and derive the encryption keys, integrity keys, and initialization vectors (IVs) from the secret key and hash values.

The SSH Authentication Protocol [15] provides user to server authencication over an established Transport Layer session. During the setup of an SSH session using the Transport Layer Protocol, an SSH client authenticates a server by verifying the server's public key provided by the SSH server. If the public key is unsigned, the client authenticates the server by matching the server's public key to a locally stored SSH host key. Alternatively, the client will trust the server's public key if the server's digital certificate is signed by a trusted CA.

SSH also supports a *leap of faith* mode. In this mode, neither prior configuration of SSH host keys nor the availability of a public key infrastructure is assumed or required. The first time a client connects with a server, the client accepts the public key presented by the server. This key is then cached for future sessions. The client essentially makes a leap of faith that no MitM attacker is present on this initial connection; if one is, the connection can be compromised. However, since the client caches the server's public key in the initial exchange, the attacker must remain present in all future SSH-TP exchanges or he will risk having the initial attack revealed when a client's attempt to connect to the true server fails.

SSH authenticates the client after the Transport Layer Protocol has authenticated the server and established a confidential, integrity protected session. Three mechanisms are supported—public key, password, and host-based. In public key mode, the client generates a signature over a message using the client's private key. The server validates the signature using the client's public key. In the password mode, the username and password of the client are sent over the encrypted SSH session for validation by the server. In the host-based mode, the client generates a signature over a message using the client's host's private key which is validated against the hostname by the server.

The Connection Protocol [16] could arguably be called the multiplexing protocol. SSH-CP multiplexes application streams over a single SSH session. Other applications using SSH are defined in separate specifications, including the use of SSH for file transfer [17].

Figure 6.7 depicts a user opening a SSH connection. The client opens a TCP connection to a server at TCP port 22. The first exchange between client and server is the INITialization message, containing protocol and software version information (here, SSH version 2.0). The client and server exchange lists of supported encryption algorithms and negotiate encryption, hash and key exchange algorithms and compression method independently for each direction of data flow. When this negotiation completes, the client and server perform key exchange (Diffie-Hellman). When the key exchange completes, both client and server share a secret K, and a hash H. Using the hash, the parties compute the server IV and client IV (when block ciphers are used) and the encryption and signing keys for both the client-to-server and server-to-client data flows.

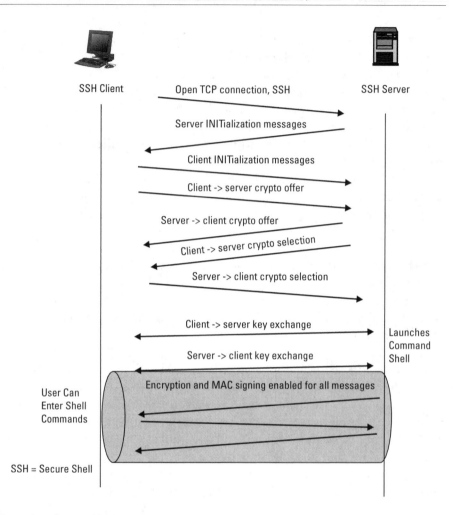

Figure 6.7   Secure Shell operation.

Client and server can now use encryption and integrity services. The server can authenticate the user over a secure channel.

The SSH standard defines both a generic port forwarding (tunneling) protocol, and also specific commands to remotely invoke a shell session and execute commands inside that shell session. So in some cases, SSH serves as a secure tunnel, and in other cases, SSH is the application protocol carried within a channel multiplex onto the SSH connection. When SSH provides a command shell session, no other application protocol is port-forwarded through the same channel as the command shell operations.

SSH could potentially be used as a SIP transport protocol; however, no current specification or implementations exist. SSH is useful in the remote management of VoIP servers and components.

6.6 Pretty Good Privacy (PGP)

Pretty good privacy (PGP) [19] was developed by Phil Zimmermann. PGP is a cryptographic system commonly used for email, file system encryption and secure archival of sensitive data. PGP uses a hybrid cryptosystem; that is, a combination of secret key and public key encryption (see Chapter 2). PGP users create public-private key pairs in a PGP and X.509 certificate formats. PGP certificates differ from X.509 certificates not only in format but by the way that authenticity is verified. As we discussed in Chapter 2, a user trusts an X.509 certificate if it trusts the issuing CA. A PGP user trusts another user's PGP certificate because he trusts other individuals who have signed the certificate. Signed PGP certificates are stored in public key repositories worldwide. PGP thus uses the concept of a "web of trust" instead of an (hierarchical) public key infrastructure. PGP key signing began as a sort of cult ceremony. For example, technical professionals meet face-to-face at IETF, and other conferences, and validate colleagues' public keys.

PGP also uses the concept of an "introducer,"someone who vouches for the identity of another. An introducer (a) signs the public key of another user (b). B can use his signed certificate with a third user (c), provided C trusts A to make valid introductions. The trusted third party characteristic of PGP introducer is similar to a CA but more general: in the PGP model, any mutually trusted party can be an introducer. Introducers facilitate the growth of the web of trust. As a web of trust begins overlap (significantly), for instance, in a community of interest, club, or private organization, a PGP user can trust and confidently use the public key of any other PGP user in that web to send a secure message, and trust that any PGP user within the web of trust who signs a message with his private key is who he claims to be.

Each PGP correspondence is an independent event. To send a private message, a PGP user generates a one time secret key which it then uses to encrypt the plaintext using a symmetric cipher. The PGP user encrypts the secret key using an asymmetric cipher and the public key of the recipient. The encrypted secret key is appended to the PGP-encrypted message. PGP encodes signed, encrypted messages in ASCII or as a MIME (multipurpose Internet mail extensions) message body part [20].

The PGP recipient receives an ASCII or MIME-encoded datum containing the encrypted message and the encrypted secret key. The secret key is encrypted using the recipient's private key, and the secret key is then used to decrypt the message. Message integrity is provided by signing a hash of the message and including it in the message as well.

PGP has two commonly used versions, 2.6.2 and 5.0. Version 2.6.2 uses IDEA as the block cipher, RSA for asymmetric encryption, and MD5 for digital signature hashes. Data transported is compressed using the ZIP or PKZIP algorithm. PGP is available as freeware for noncommerical use. Commercial versions of PGP offer many enterprise-grade features. In addition to e-mail security, PGP can be used to secure instant messaging, local files, and archive files.

Figure 6.8 illustrates the operation of PGP when it is used to secure e-mail (conceptual).

6.7 DNS Security (DNSSEC)

Domain name services security extensions (DNSSEC) [21] adds origin authentication of DNS data, integrity protection, and authenticated denial of existence to DNS [22]. Origin authentication means that a security-aware DNS resolver has a reliable means of acquiring a zone public key, and can authenticate that

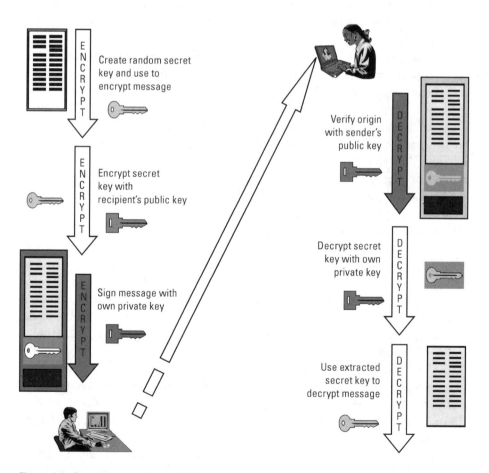

Figure 6.8 Securing e-mail using PGP.

zone's signed data by decrypting the signed hash of that zone's data. Authenticated denial of existence provides a way for DNS resolvers to trust that a negative response to a DNS query is correct. DNS resolvers can query a new DNS resource record to acquire the public key of a zone signing key pair.

DNSSEC is concerned with the authentication and integrity of the DNS records. It does not provide confidentiality for DNS transactions. In the DNSSEC model, the person who requests DNS records and the person who returns the results is unimportant. What is important is that the responses, both positive and negative, are identical (correct and complete) to the resource record maintained by the authoritative DNS server of the queried domain.

To support DNSSEC, DNSSEC resolvers must support extension mechanisms for DNS, (EDNSO, [23]). EDNSO extends the message header to accommodate additional labels, flags, and RCODES, and accommodates DNS messages of length greater than 512 octets.

The basic components of DNS are shown in Figure 6.9. The DNS resolver NS requests from an application and queries servers known as name servers. Due to the delegation model of DNS, several name servers may need to be queried.

DNSSEC adds four new DNS resource records (RR), which are listed in Table 6.2.

The RRSIG resource record signature RR contains a digital signature for a particular RR. In a secure DNS zone, there will one RRSIG for each resource

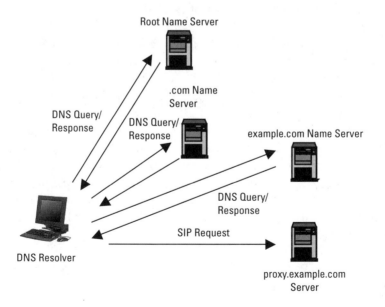

Figure 6.9 DNS components.

Table 6.2
DNSSEC Resource Records

Record	Name
RRSIG	Resource record signature
DNSKEY	DNS public key
DS	Delegation signer
NSEC	Next secure record

record. The public key associated with the private key used to sign the RRSIG can be manually configured in the DNS client or it can be learned through the DNS public key (DNSKEY) and resource records. If a DNSKEY is public key that the DNS resolver has authenticated, then the RR information can be validated. If it is not, the delegation signer (DS) record is used to point to the next higher level of delegation DNSKEY record. This can be used to go up the hierarchy until an authenticated key is returned. At this point, the RR information can be validated. At a minimum, a DNSSEC-capable resolver must have the public key of the top level of the DNS (root), as every DNSSEC DS chain will ultimately refer up to the root.

The Next Secure Record (NSEC) is used to securely validate negative response, that is, the absence of a RR. The NS RR is used to link every valid RR in the domain. By retrieving the set of NS RRs, a secure resolver can securely verify the absence of a record.

A secure DNS resolver indicates that it wishes to receive DNSSEC records by setting the DO bit. A secure DNS name server recognizes the DO bit and responds with the appropriate DNSSEC RRs. Note that all DNS elements in a chain must support DNSSEC, not just the resolver and name server.

VoIP systems often make use of DNS queries. For example, DNS can be used to resolve a host to a given IP address using A (address) records. Or, DNS can be used to locate SIP servers in a domain using SRV (service) records. DNS can also be used to determine what transport protocols are supported by a given SIP server using NAPTR (Naming Authority Pointer) records. In Chapter 11, we will see how DNS NAPTR records can also be used to map a telephone number to a URI using a protocol known as ENUM. Any DNS query can be secured using DNSSEC.

References

[1] Phifer, L., *VPNs: Virtually Anything?* A Core Competence Industry Report, http://www.corecom.com/html/vpn.html.

[2] Kent, S., and K. Seo, "Security Architecture for the Internet Protocol," RFC 4301, December 2005.

[3] Kent, S., "IP Authentication Header," RFC 4302, December 2005.

[4] Kent, S., "IP Encapsulating Security Payload," RFC 4303, December 2005.

[5] Harkins, D., and D. Carrel., "The Internet Key Exchange," RFC 2409, IETF, November 1998.

[6] Beaulieu, S., and R. Pereira, "Extended Authentication Within IKE (XAUTH)," http://www.vpnc.org/ietf-xauth/draft-beaulieu-ike-xauth-02.txt.

[7] Dierks, T., and C. Allen, "The TLS Protocol Version 1.0," RFC 2246, January 1999.

[8] Dierks, T., and E. Rescorla, "The TLS Protocol Version 1.1," June 2005, IETF Internet-Draft, http://www.ietf.org/internet-drafts/draft-ietf-tls-rfc2246-bis-13.txt.

[9] Hollenbeck, S., "Transport Layer Security Protocol Compression Methods," RFC 3749.

[10] Friend, R., "Transport Layer Security (TLS) Protocol Compression Using Lempel-Ziv-Stac (LZS)," RFC 3943, November 2004.

[11] Rescorla, E., and N. Modadugu, "Datagram Transport Layer Security," June 2004, IETF Internet-Draft, www.ietf.org/internet-drafts/draft-rescorla-dtls-05.txt.

[12] Postel, J., "User Datagram Protocol," RFC 768, August 1980.

[13] Ylonen, T., and C. Lonvick, "SSH Protocol Architecture," RFC 4251, January 2006.

[14] Ylonen, T., and C. Lonvick, "SSH Transport Layer Protocol," RFC 4253, January 2006.

[15] Ylonen, T., and C. Lonvick, "SSH Authentication Protocol," RFC 4252, January 2006.

[16] Ylonen, T., and C. Lonvick, "SSH Connection Protocol," RFC 4254, January 2006.

[17] Galbraith, J., O. Saarenmaa, T. Ylonen, and S. Lehtinen, "SSH File Transfer Protocol," IETF Internet-Draft, work in progress, June 2005.

[18] Deutsch, P., and J.L. Gailly, "ZLIB Compressed Data Format Specification Version 3.3," RFC 1950, May 1996.

[19] Atkins, D., W. Stallings, and P. Zimmermann, "PGP Message Exchange Formats," IETF RFC 1991, August 1996.

[20] Elkins, M., "MIME Security with Pretty Good Privacy (PGP)," RFC 2015, October 1996.

[21] Arends, R., R. Austein, M. Larson, D. Massey, and S. Rose, "DNS Security Introduction and Requirements," RFC 4033, March 2005.

[22] Mockapetris, P., "Domain Names—Implementation and Specification," STD 13, RFC 1035, November 1987.

[23] Vixie, P., "Extension Mechanisms for DNS (EDNSO)," IETF RFC 2671, August 1999.

7

General Client and Server Security Principles

7.1 Introduction

This chapter will discuss various client and server security principles which can be applied to secure VoIP systems and networks. Since VoIP systems typically rely on communication with hosts on the public Internet, or communication outside an enterprise to a service provider, these general security requirements are quite important.

Well-designed client and server security implementations assume that external security measures, such as firewalls, VPNs, network IDS, and IPS, are not sufficient, but that layers of security should be applied wherever possible. Apply a defense in depth to complement firewall and security gateways by securing endpoint devices and servers.

Today, experts recommend that organizations should build security "inside out." Rather than thinking about perimeter security first, and complementing the perimeter with additional measures "behind the firewall," many organizations now secure individual systems so that they are resilient to, or hardened against, insider attacks. Increasingly, organizations are applying isolation, segmentation and compartmentalization techniques. Watertight hatches on submarines can contain flooding within a compartment; intelligently applied system and application segmentation can prevent a successful attack against one system from spreading to others.

Client security is critically important in VoIP because VoIP technology is likely to become very mobile. Server security is equally critical. Even though servers may be protected by firewalls, the real-time nature of VoIP service puts a premium on availability. This premium will only increase as VoIP supplants PSTN and cellular services in the future.

7.2 Physical Security

Physical security requirements vary from organization to organization. Organizations small and large acknowledge and provide physical security to prevent tampering and destruction of critical systems and data; how each implements such measures depends on budget constraints and what risk the organization will tolerate. Security measures may be influenced by regulatory guidelines based on location, and industry, to name a few. While security measures may vary, the threats organizations must consider are common.

Some of these are:

- Unauthorized entry;
- Tampering;
- Theft;
- Destruction resulting from human acts (malicious and accidental);
- Destruction resulting from natural disasters;
- Loss of power, power surges;
- Substitution;
- Radio, electrical, magnetic interference.

Physical security and information security are complementary. As VoIP is an IT application, we focus primarily on information security measures in this book.

7.3 System Security

An application is only as secure as its operating system, so, as a result, it is critical to first secure the operating system. Most VoIP applications run on commercial operating systems, and no commercial operating systems can claim to be vulnerability and exploit free. Many "best practices" have emerged for securing both client and server operating systems.

7.3.1 Server Security

Volumes have been written regarding the best ways to secure or "harden" the configuration of servers for both Windows and UNIX-based servers. A partial list of common practices for securing servers, regardless of OS type, follows. For more exhaustive descriptions of ways to harden operating systems, see [1–5].

Maintain current patch levels for the operating system and any services the system supports (DNS, DHCP, active directory). The vast majority of successful attacks involved reported exploits and known vulnerabilities for which patches were available but had not been installed.

Only run necessary services. Run services using the most secure configuration possible, the strongest authentication available, and the most stringent access permissions possible. For each required service, disable unnecessary features and make certain only authorized ports are in a listening/responding state. Enable service level auditing (DNS and web logging). Restrict service operating mode (limit web server access to system commands). Use file permissions to restrict the content types of accessible to services. Remove or prohibit access to sample and test directories, scripts, and executables.

Review and eliminate all default configuration settings. Delete or disable default accounts and reset default passwords at the OS and application levels. Disable any preinstalled vendor remote assistance accounts if you will not use vendor support channels.

Maintain rigorous control over user accounts. Only create such accounts as are absolutely necessary. Do not create accounts for users that result in access privileges on systems they have no business reason to access. Disable and archive user accounts quickly following employee termination, extended leave, hospitalization, or other circumstance where the user is not expected to require access. Only grant administrative privileges to parties who are authorized system administrators.

Apply stringent security policies to user accounts and permit only such user rights as are necessary to conduct authorized activities. Where necessary and possible, compose group memberships along "need to know" rather than organizational lines. Follow the Law of Least Privilege when configuring access permissions on applications, operations within applications, file systems, and individual files. Audit account login, account management, and privileged access events (at a minimum). Increase audit levels for any activities that are deemed sensitive, as, for example, access to system configurations.

Restrict removable media. Removable media can be used for theft of information and tampering. Limit the types of removable media that may be connected to servers (for example, disable USB device connection). Where appropriate, restrict content types that can be stored on removable media.

Restrict the execution of applications from removable media to prevent (auto) installation of malicious code. Consider using only approved media that can be authenticated. Consider the use of data encryption and integrity measures on removable media to prevent theft.

Antivirus protection on servers is often neglected, but it is especially important in homogeneous (Microsoft) environments or "monocultures," where malicious code can execute on client and server versions of an OS.

Antitampering and host intrusion detection on servers provide an additional layer of defense on top of an already hardened OS. These measures can alert system administrators to any unauthorized modification, substitution or addition of configuration and policy files as well as files and libraries that are essential to the correct (and secure) operation of a server.

Perform routine data archival. Include configuration data for all (VoIP) applications and security measures. Use secure erase methods for file deletion.

The Center for Internet Security (CIS, [6]) offers benchmarking tools and policy templates for all major commercial operating systems, popular network devices, and commonly used server applications (Apache, Oracle, Exchange). Organizations can use these to assess and implement a resilient, secure server environment for VoIP and data systems.

7.3.2 Client OS Security

Today, mobile data devices do not remain behind corporate Internet firewalls and security gateways. VoIP devices will be even more mobile than data handhelds and laptops and as vulnerable to attack if not more so. Client devices are more likely to be affected by attacks that leverage human interaction such as phishing, "shoulder-surfing" and social engineering. They are more susceptible to misconfiguration and malicious code. Client devices can provide dangerous exploit vectors to servers: a compromised client device can provide an attacker with a secure tunnel to an organization's internal network.

Common practices for securing data handhelds and laptops can be applied to many VoIP endpoints. Many of these measures are similar to those we have prescribed for server security, and include the following:

Maintain OS and application security patch levels. This measure can be more difficult to implement on client devices than servers because of the sheer numbers and increased mobility of client devices. Centralized patch management applications and network admission control can be implemented to help address this problem.

Run only necessary client services and approved, licensed applications. Organizations that permit unrestricted download and installation of software, plug-ins, and even ring tones have a higher security incident rate than ones that limit user self-administration, and are more exposed to copyright infringement and other licensing issues.

Prohibit client devices from enabling services. There are few business reasons for users to host web, FTP, telnet, routing, and other infrastructure services. These are unlikely to be operated in the same secure fashion as systems administered by IT and pose serious threats.

Restrict type and use of removable media. Organizations may not require the same restrictive policy governing removable media for client devices as they impose on servers, but many organizations would benefit from a policy that provides some enforcement and auditing of removable media.

Restrict user access to policy and configuration files. Policy is difficult to enforce when users have local administrative privileges and are able to modify any client device configuration that alters or disables security measures. Limiting user self-administration and using central policy definition (as in group security policy and distribution via Microsoft's Active Directory) help minimize exposure from users who "dumb down" endpoint security as a remedy for a temporary problem.

Enable auditing and restrict user access to event logs. Event auditing on client devices is a useful practice. Analysis of client device event logs can help identify configuration problems and rogue user activities. Logs can also provide security staff with early warnings of possible attacks; if protected against modification or deletion, they provide useful information to help reconstruct an attack following a security incident.

Layer client endpoint defenses. Client devices should have antivirus and antispyware protection. These should be configured to perform routine scans and to update malware definitions regularly. Personal firewalls and intrusion detection should be used even when endpoint devices are configured securely. Configure personal firewalls with the strictest inbound and outbound policies possible. By using per interface policies, client devices can use personal firewalls effectively whether they are connecting to public, home, or work LAN/WLAN environments.Use logging features offered by personal firewalls to complement event logging. If the organization uses a VPN that requires client configuration, such as an IPSec network adapter, avoid split-tunneling, define the strongest authentication, encryption and authorization policy possible, and protect the configuration from user modification.

CIS offers benchmarking tools and policy templates for client operating systems (Windows XP Professional) as well. The templates are particularly useful for organizations that will use VoIP soft phones and want to make use of group security policy administration through the active directory.

The market for VoIP hard phones is still relatively immature, and some security measures may not be available in early models today. Consumer grade VoIP hard phones are unlikely to have advanced security features. When evaluating VoIP hard phones, look for a phone that has configurable security features and harden the phone as you would a data device. Choose phones that provide secure administrative interfaces (many consumer phones are configured through an un-secured HTTP interfaces). Require password/PIN authentication for all administrative access. Run a port scan against the phone to learn what services are responding; investigate what applications are listening, and what they do. Disable any nonessential service.

7.4 LAN Security

Every organization should secure its internal LAN infrastructure. Several implementation strategies used on data networks today can be applied to VoIP networks:

Control admission to networks. IEEE 802.1x-based device authentication, network admission control [7], network access protection [8], and similar emerging policy enforcement measures enhance the security of LANs by preventing stations that do not meet security configuration criteria from connecting to LANs and WLANs (see Sections 7.4.1 and 7.4.2).

Control physical LAN connections. Use managed Ethernet switches and make use of administrative controls to disable unused Ethernet ports. Where possible, identify LAN MAC addresses that are authorized to connect to physical LAN ports. Monitor radio spectrum to identify rogue access points, interference from neighboring WLAN environments, and to detect radio spectrum attacks.

Use virtual LANs or physical network segmentation to separate data, voice, and "critical infrastructure" services, to protect each segment from "leak over" attacks. Run infrastructure services on separate systems, to prevent a successful attack on one system from threatening the entire service infrastructure. To illustrate why this is a "best practice," consider two scenarios. In one scenario, an organization runs a voice mail application and SIP proxy server software on the same physical server. If the voice mail application is compromised and the attacker gains administrative privileges on this server, he potentially "owns" the SIP proxy server as well. In the second scenario, the voice mail application is run on separate hardware from the SIP proxy server. If the voice mail system is compromised, the attacker must launch a separate, successful attack against the system hosting the SIP proxy server (see Section 7.4.3).

Apply the security principle "defense in depth" by adding security measures at strategic points inside the internal network. Strategic deployment of internal firewalls and intrusion detection systems not only provide additional layers of defense, but they move defenses closer to assets, where security policies can be tailored to the specific application mix permitted on individual segments (see Section 7.4.4).

7.4.1 Policy-Based Network Admission

Endpoint and admission control techniques are the virtual corollaries of immigrations and naturalization agencies. In many countries, an individual must present more than a valid passport before he is allowed entry. For example, the individual may have to satisfy visa and immunization criteria. These measures are implemented to protect the citizenry of the country from attack and disease. The corollary in the network security world is simple: user and device authentication alone are not sufficient criteria to grant access to a network. Additional security criteria must be satisfied.

Policy-based admission control is part of a security framework not a technology. Admission policies set conformance criteria for client devices. There are many criteria an organization might check before admitting a client device, including:

- Antimalware measures and personal firewalls must be installed, executing, and configured properly.

- Tested and approved patch and hot fix levels must be installed.

- Local security policy and VPN adapter configuration must be current and consistent with policy the organization intends to enforce.

- No unlicensed or extraneous software, drivers, dynamic libraries, or other forms of executable software are installed.

- Only authorized processes are executing.

Technologies used to enforce network admission policy can be configured to deny access to client devices that do not satisfy admission criteria. Nonconforming devices can also be redirected to a remediation server, where an authenticated user can download and install missing security components and update configurations. Organizations may also choose to grant limited access to nonconforming devices, so that a guest may access the Internet but no internal systems. Organizations can also grant reduced access to allow mobile employees to perform a limited set of tasks without placing internal assets at risk. The employee might then defer the process of bringing his device to compliance until he has access to broadband or software media.

Figure 7.1 depicts a generic model of policy-based network admission.

Most current network admission solutions require a client-side agent. The agent performs the "scan before connect" that assesses whether the client device is policy conforming. Initially, such agents may not be available for all VoIP endpoints; however, just as antivirus measures are more available for data handhelds today than when the virus threat to handhelds was first demonstrated, client agents are likely to appear as policy-based network admission solutions are more widely adopted.

7.4.2 Endpoint Control

Endpoint control is a related policy-based enforcement mechanism. In the data world, endpoint control (EPC) assumes that user may access the company network from non-work systems. This is common today with data applications. Employees use non-work computers in their homes and at business centers in hotels to access company intranets. In the future, employees may use nonwork VoIP phones in similar circumstances. The problem in such scenarios is that an organization's IT may not be able to install resident admission control software on a non-work client device. EPC works around this problem by employing

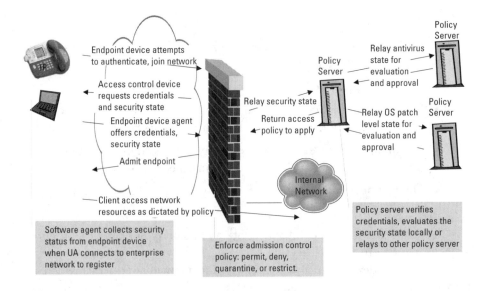

Figure 7.1 Generic model for policy-based network admission.

temporary (Java or ActiveX) software agents. EPC agents can scan on connect for many of the same security criteria as policy-based network admission solutions.

Most SSL/TLS VPN vendors offer some form of EPC. In the current forms, EPCs often perform application-centric checks and can be configured to restrict the set of applications a user may access from uncontrolled systems (prohibit FTP access from any nonwork system) and restrict specific commands a user may perform from uncontrolled systems (prohibit use of FTP GET operation from a non-work system). EPCs can also use endpoint location and user identity authentication as the bases for authorization decisions to quarantine or limit access when endpoint does not warrant full trust.

Endpoint control also provides exit control. The goal of exit control is to assure that nothing sensitive is left on a nonwork client device when a user concludes his session. For browser based applications, this means that cached credentials, hyperlink histories, temporary files, spooled printer files, and local copies of sensitive files are removed from the nonwork device when a user session is terminated.

7.4.3 LAN Segmentation Strategies

Segmentation strategies can measurably improve internal LAN security. One strategy segment servers according to services and information they host, as in managing separate VLANs or LAN segments for:

- Voice infrastructure LAN(s);

- Application LAN(s) and data centers;

- Infrastructure services LAN(s), for instance, separate segments for DHCP, DNS, authentication/identity, directory, accounting servers;

- Administration and log server hosts.

Another strategy calls for the separation of clients from servers. Organizations with operating system "monocultures" find this form of segmentation may insulate servers from a rapid spread of malicious code from infected client devices.

Yet another segmentation strategy calls for separation of data and voice clients. The theory here is that an integrated services network doesn't imply that the network topology must integrate data and voice devices. Some organizations may find it easier to maintain quality of service for VoIP systems when they are isolated from bursty data applications. Security policies may also be simpler to implement at security devices when permissions can be defined for voice-only and data-only segments.

Figure 7.2 illustrates a generic model for LAN segmentation, using a VLAN approach. Segmentation can also be performed in this manner using a switch for each segment and trunking or interconnecting the switches in a hierarchical or mesh topology.

7.4.4   LAN Segmentation and Defense in Depth

The term firewall is almost always interpreted to mean *Internet-facing* firewall. Internet firewalls separate a collection of internal or trusted networks from the generally-not-to-be trusted Internet. This is simply a policy assertion that characterizes the role an Internet firewall performs. Generally speaking, a firewall separates two or more security policy domains. Only traffic that is permitted by policy may cross from one policy domain to another. The broader definition accommodates applications of firewalls as a means of providing compartmentalization [9]. Firewalls can be used to restrict the flow of traffic between network segments according to the services the segments provide (data and voice), business unit (marketing and engineering), or information sensitivity (personal medical information databases from product documentation). Firewalls provide effective means of asserting traffic policy when LAN segmentation is used. In VoIP-enabled networks, interdepartmental or compartmentalizing firewalls can be used to enforce policy and protect VoIP server farms, and Internet and intranet data centers. Figure 7.3 illustrates several roles firewalls can play when services are compartmentalized.

When only Internet firewalls are present, the security policy must accommodate application needs for all users and applications, for clients and servers alike. When firewalls are moved closer to servers, organizations can refine policy.

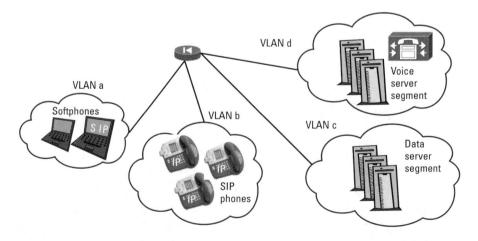

Figure 7.2   LAN segmentation (using VLANs).

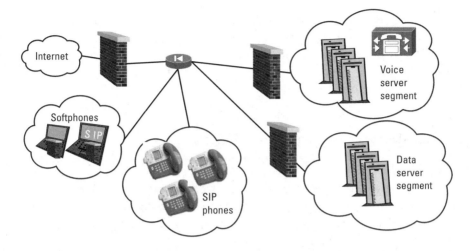

Figure 7.3 Firewall deployment.

In a voice-data network, a firewall placed in front of VoIP servers can be configured to permit only protocols needed to support voice service. The firewall can restrict access to only the set of ports needed to operate the VoIP service be opened to servers. For example, if a server protected by the firewall is functioning as a SIP proxy server, the firewall is configured to allow traffic to ports 5060 and 5061. Ports that might permit data services (telnet, rlogin, SNMP) are disabled and blocked by the firewall. This further insulates VoIP servers from attacks from internal hosts that use data applications.

Organizations don't have to invest in stand-alone systems to add layers of firewalls. Organizations can use server firewall software, firewall network adapters, and firewall-capable LAN switches to build a defense in depth.

7.5 Secure Administration

Remote administration is not merely a convenience, but essential. Most organizations need remote administration to provide 24/7 operations, support, and emergency response. All remote administration should be secured by using security protocols that provide authentication, confidentiality, and integrity, and (if possible) from trusted systems. Insecure remote access methods such as FTP and telnet should be avoided, even when remote access is performed over internal networks. Secure shell [10], secure file transfer, secure web administration using HTTP/SSL, are available for most commercial OSs and are implemented on most enterprise-grade VoIP (and data) systems. Be suspect of VoIP systems that cannot provide these or an equivalent proprietary secure administration channel.

Most organizations require multifactor authentication for remote administration. The expense of a two-factor authentication system for a small population of system and security administrators is minor compared to the damage an attacker can inflict if he gains administrative privileges on servers that are critical to business operations.

7.6 Real-Time Monitoring of VoIP Activity

Real-time monitoring of VoIP activity is key to detecting a number of attacks. Fraud can be detected and thwarted at early stages by analyzing calling patterns. For example, some real time monitoring systems can identify account abuse such as a VoIP user who pays consumer VoIP flat rate service but uses the service to operate a pay phone business.

The use of intrusion detection systems (IDS) can also benefit VoIP networks. Even a data-centric IDS can be useful in detecting DoS, DDos, attacks against TCP/IP, and OS-specific exploits. Intrusion detection and analysis cannot always be performed on data from a single source. Traffic data from a number of points on the network may need to be collected and analyzed together, possibly using an additional security information management (SIM) application. Even if such an analysis system is not available, the data to enable this type of system should be collected and archived at the least. The archived logs will provide the ability to do after the fact analysis of an attack, at the very least.

7.7 Federation Security

The federation, or interconnection, of VoIP systems from different service providers or enterprises poses significant security risks. However, the advantages of doing so with respect to cost savings and feature benefits make this a logical next step for VoIP systems. As such, federation should only be done when appropriate security mechanisms are in place. At a minimum, all federated VoIP exchanges should be authenticated, even if it is only validating a certificate. To federate without even this basic identity requirement is to open VoIP up to e-mail-like spam and abuse, as described in Chapter 13.

7.8 Summary

This chapter has listed a number of general client and server security principles. VoIP is yet another data application, and nearly all industry best practices for securing data networks are applicable to integrated voice-data networks. VoIP does introduce a number of unique security challenges. Because VoIP is still a relatively new application, methods for mitigating a number of voice-specific threats may not be easily implemented, but by following the recommendations we describe, organizations should be able to significantly reduce risk when introducing VoIP.

References

[1] Bragg, R., *Windows Server 2003 Security: A Technical Reference*, Reading, MA: Addison-Wesley Professional Series, 2005.

[2] Deseglio, M., *Securing Windows Server 2003*, 1st ed., Sebastopol, CA: O'Reilly Media, 2004.

[3] Bauer, M., *Linux Server Security,* 2nd ed., O'Reilly Media, 2005.

[4] Daniel J. Barrett, Richard E. Silverman, and Robert G. Byrnes, *Linux Security Cookbook,* 1st ed., Sebastopol, CA: O'Reilly Media, 2003.

[5] Cox, P., and T. Sheldon, *Windows 2000 Security Handbook,* 1st ed., New York: Osborne/McGraw-Hill, 2000.

[6] The Center for Internet Security, http://www.cisecurity.org.

[7] http://www.cisco.com/warp/public/cc/so/neso/sqso/csdni_wp.htm.

[8] Windows Server 2003: Network Access Protection, http://www.microsoft.com/windowsserver2003/technologies/networking/nap/default.mspx.

[9] Piscitello, D., "Interdepartmental Firewalls: Where to Put Them and Why," http://www.corecom.com/external/livesecurity/firewallplace.html.

[10] Barrett, D., and R.E. Silverman, *SSH, The Secure Shell: The Definitive Guide,* 1st ed., Sebastopol, CA: O'Reilly, 2001.

8

Authentication

8.1 Introduction

In Chapter 5, we introduced authentication as a process where an entity proves its identity by presenting credentials that are difficult for anyone but the actual identity to produce. Authentication is the first or enabling measure for VoIP security and in general any Internet security policy enforcement mechanism. Through authentication we corroborate that an individual is who he claims to be, or that a system is expected and rightful recipient for a given communication, or that an application may rightfully process a request or transaction on a user's behalf.

Many authentication methods challenge an entity (user, client endpoint device, server) to prove its identity, and are referred to as challenge-response systems. The image of an armed guard calling, "Who goes there?" from an old war movie accurately depicts a challenge-response mechanism based on a group shared secret password. Every day, thousands of people use two-factor authentication when they withdraw cash from an ATM (the ATM card is a "token factor" and the PIN is a second "secret factor" known only to the account holder and financial institution).

We often think of authentication in terms of a user or endpoint device authenticating to a security gateway, a web application, or a SIP proxy server. Many of the same authentication methods are used by servers to authenticate peer servers. For example, in Chapter 6, we explained how the security protocol IPSec performs endpoint authentication using the IKE protocol. And in Chapter 11, we will explain a method of securing SIP signaling that chains TLS sessions across multiple SIP proxy servers.

Organizations employ authentication methods to enable security services in a variety of applications, as for example:

- Endpoint device access—PC OSs challenge users to present logon credentials, handhelds and VoIP phones challenge users for PINs.

- Application access—applications and proxies challenge local and remote users to present account names and credentials (commonly passwords and tokens) before they are allowed access.

- Network admission—security systems challenge endpoints to present digital credentials (preshared keys, digital certificates, unsigned public keys) before they permit a device to connect to a network.

- Network interconnection—security gateways and proxies challenge endpoints or peer servers to present digital credentials (preshared keys, digital certificates, unsigned public keys) before they agree to establish secure tunnels from an endpoint or peer server to networks they protect.

- Secure administration—management applications and systems challenge users to present credentials and prove they have administrative privileges before providing access to application and device configurations.

A critical component of any authentication scheme is the identity or account database or directory. Conceptually, this is a secure repository of all identities, credentials, and access privileges. In practice, organizations often have many identity and user account databases. Some may be specific for e-mail (as Lotus Notes Address Book) or secure access (a SecurID or RADIUS authentication database) and some may be for network application access (Microsoft's Active Directory).

Any security system that performs authentication must maintain a local user account database or use a special-purpose *authentication protocol* to query an external authentication database or directory. In certain implementations, security systems may not participate in authentication, but will proxy and forward authentication protocol exchanges between a user and an *authentication server*. Where authentication requires access to external databases and servers, measures must be taken to protect these exchanges from attacks, such as by using security protocols or compartmentalization and traffic isolation.

This chapter describes two authentication methods and protocols that VoIP network operators are likely to employ. Many standards-based and proprietary authentication protocols exist, and the examples in this chapter are representative of how authentication protocols generally operate.

8.2 Port-Based Network Access Control (IEEE 802.1x)

Port level access control based on MAC (Media Access Control) address filtering is often used as part of a LAN access control strategy. MAC address filtering provides a form of authentication and access control, and is based on the original implementation and manufacture of Ethernet equipment, where 48-bit Ethernet addresses were hard-wired to adapters, creating a permanently bound relationship between an identity and an entity. Today, LAN adapter configuration is more flexible. Users, and thus, attackers can configure a LAN adapter with and impersonate any MAC address. If an attacker can gain physical access to a LAN or WLAN, and traffic is exchanged without encryption, he can capture traffic, collect MAC addresses that are presumably permitted access, and use these to subvert this extremely weak form of authentication.

IEEE 802.1x [1] is a port-based access control standard that enables authentication and key management for IEEE 802 networks. It is based on the Extensible Authentication Protocol [2], a PPP extension. Adaptations of EAP accommodate different authentication methods, including EAP-MD5, EAP-TLS [3], and EAP-TTLS [4]. IEEE 802.1x, is not exclusively for wireless LANs and can be used with any IEEE 802 LAN physical medium. IEEE 802.1x/EAP is alternatively referred to by the acronyms EOPOL (EAP Over IEEE 802.3 LANs) or EAPOW (EAP Over IEEE 802.11 WLANs).

EAP messages are encapsulated directly in 802.1x802.1x messages, with no additional overhead. IEEE 802.1x authentication for wireless LANs has three main components:

- Supplicant (client software);
- Authenticator (access point);
- Authentication server (a RADIUS/AAA server).

The relationships between these components are illustrated Figure 8.1. Conceptually, a client endpoint associates with an IEEE 802.11 access point that supports IEEE 802.1x/EAP. The access point allows the client to connect, but the client can only exchange EAP traffic with the access point at this time. In an IEEE 802.3 LAN environment, the client is only able to exchange EAP traffic with the LAN switch. The access point acts as a proxy agent to an authentication server and relays EAP authentication messages between the client and authentication server. If the authentication server accepts the client's credentials, it tells the access point (LAN switch) to admit the client to the WLAN (LAN).

Figure 8.2 illustrates an EAPOL message exchange. As shown, a LAN station (supplicant) connects to a LAN switch (authenticator). The LAN switch detects the client and enables the station's LAN port, sets the port state to blocked, and silently discards all traffic from the LAN station except 802.1x

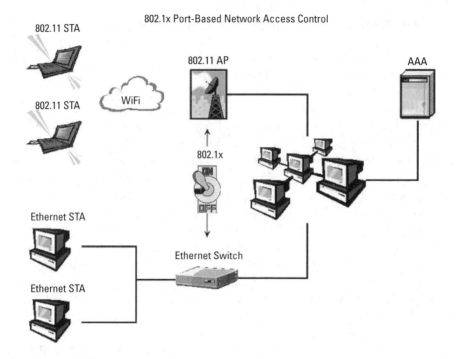

Figure 8.1 IEEE 802.1x/EAP components.

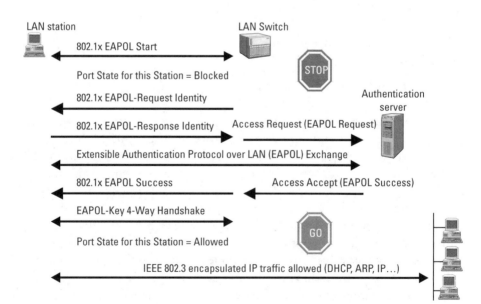

Figure 8.2 IEEE 802.1x/EAP message exchange.

messages. When a LAN station detects physical signal, it transmits EAPOL-START messages. The LAN switch replies with an EAPOL-REQUEST IDENTITY message (in some implementations, LAN switches send EAPOL-REQUEST IDENTITY messages without waiting for EAPOL-START). The LAN station submits its identity information for authentication in an EAPOL-RESPONSE message. The LAN switch forwards the supplicant's EAPOL-response packet to the authentication server. The authentication server attempts to verify the supplicant's credentials using the authentication algorithm implemented on this LAN (EAP-TLS, EAP-MD5, EAP-TTLS...). If the authentication server verifies the supplicant's credentials, it returns an ACCEPT message to the LAN switch, otherwise the server returns a REJECT message. The LAN switch reacts to the EAPOL ACCEPT response by sending an EAPOL SUCCESS message to the LAN station, and changing the LAN port for this station to an allowed state. Unless additional admission controls must be satisfied, the LAN station can proceed to use the LAN. If the authentication server returns a EAPOL REJECT message, the LAN switch issues an EAPOL FAILURE response to the client, leaves the LAN port for this station in the blocked state, and continues to silently discard traffic other than EAPOL messages.

More than 40 authentication options exist for IEEE 802.1x/EAP. EAP-MD5 uses a simple MD5 hash of the shared secret for authentication. The hash is not signed or protected from a MitM or eavesdropper. In 802.11 wireless networks, this is an important consideration. The EAP-TLS [3] approach utilizes TLS transport between the end device and the access point. This approach uses client certificates to authenticate the user. The issue of managing end user certificates will be discussed in Chapter 12. The EAP-tunneled TLS (EAP-TTLS) [4] approach uses TLS transport but uses a shared secret as the authentication mechanism. Protected EAP (PEAP) provides secure mutual authentication and legacy sub-authentication [5]. Lightweight EAP (LEAP) provides mutual authentication based on password challenge-response [6]. EAP-FAST (fast authentication via secure tunneling) tries to overcome the problems associated with roaming across multiple WLAN access points by using shared secret keys to speed up 802.1x reauthentication [7].

Table 8.1 compares the security services and features of popular EAP types.

Once IEEE 802.1x/EAP processing is completed, port-based access filtering can be used to enforce security policies with some confidence. Additional admission criteria can be imposed on the client as well; for example, the endpoint device may have to demonstrate that it has implemented a baseline level or security before being allowed to access the network (see Chapter 5).

IEEE 802.1x/EAP can be used in VoIP networks to prevent unauthorized endpoints from connecting to LANs and WLANs. Depending on how identities are managed, IEEE 802.1x can be used to restrict endpoints to specific LAN segments and WLANs. For example, an organization can create a security policy

Table 8.1
Comparison of EAP Types

	EAP-MD5	EAP-TLS	EAP-TTLS	Protected EAP	Lightweight EAP	EAP-FAST
Server authentication methods	None	Certificate	Certificate	Certificate	Password hash	Certificate or shared secret
Supplicant authentication methods	Password	Certificate	PAP CHAP MS-CHAPv2 generic token	Other EAPs		Secret plus MS-CHAPv2
Dynamic key delivery	No	Yes	Yes	Yes	Yes	Yes
Vulnerability	Very high	Very low	Low	Low	High	Medium
Risks	Identify exposure, hash cracking	Identify exposure	Possible identify exposure MitM risks	Possible identify exposure MitM risks	Identify exposure, hash cracking	Possible auth server spoofing

that confines VoIP applications to a single LAN segment and enforce that policy using IEEE 802.1x to authenticate only those devices that can submit "voice device" credentials.

8.3   Remote Authentication Dial-In User Service

The Remote Authentication Dial-In User Service (RADIUS) protocol provides what are commonly described as the "triple-A" services: authentication, authorization, and accounting. RADIUS is used in private networks and by Internet and broadband access providers to transport user account credentials from access servers, VPN concentrators, access points, firewalls and proxies to an authentication server for processing. RADIUS is also used to collect accounting information from access servers for auditing and billing purposes [8] and to pass connection configuration and authorization information for authenticated users to access servers.

Although RADIUS is a client-server protocol, the RADIUS client is not a user endpoint device but an access server that proxies an authentication method between a user and an authentication server. For example, when a web-based transaction requires user authentication, the web server often presents a dialog box to a user requesting user account and password information. This exchange is often performed over a TLS connection between a user's browser and the web server. The web server receives the user's credentials and forwards them to an authentication server using the RADIUS protocol (and hence is the client in this exchange). The server authenticates the user, and returns a positive or negative response to the web server, which in turn allows the user to access the desired hyperlinks or blocks subsequent access. This proxy behavior is similar to the behavior of the LAN switch in our IEEE 802.1x/EAPOL example in section 8.2. RADIUS clients can also authenticate client endpoint devices that use link framing protocols such as PPP. In such cases, the link framing protocol conveys authentication information (as in PPP-CHAP), and the RADIUS client must extract these credentials before forwarding them to an authentication server using the RADIUS protocol.

RADIUS messages consist of a header and set of attributes that vary according to the message type:

- *Access-Request* messages are sent by RADIUS clients to RADIUS authentication servers to request authentication and authorization services. Depending on the authentication method employed, they contain such attributes as User-Name and User-Password, type of service requested login or link framed protocol).

- *Access-Accept* messages are positive responses to an Access-Request. Attributes in a response are influenced by service type. For example, different routing and addressing attributes are returned to an access server when a broadband router is authenticated depending on whether the router is configured with a static or dynamic IP address, and whether the router forwards packets to other public IP networks.

- *Access-Challenge* messages support multifactor authentication methods. This message typically contains a challenge the access server must present to the user, who in turn must generate an appropriate response. For example, in a one-time password system, the user must respond to the challenge with a time-synchronized numeric code displayed on his token card plus his secret PIN. Only the holder of the token and the server can know the combination of token code and PIN at any given time, and the combination can only be used once.

- *Access-Reject* messages are negative responses, indicating that the authentication server judged one or more attributes in an Access-Request unacceptable.

Many security systems (application proxy firewalls, SSL VPN proxies, SIP proxies) can enforce security policies on individual users and user groups. Such systems often act as clients and call upon RADIUS servers to handle authentication requests. Consider a situation where an organization defines a security policy that limits access to a file server to authenticated members of a group called engineering. Security administrators create accounts at a RADIUS-capable authentication server for members of this group. If a user attempts a file transfer to this file server, a firewall or SSL proxy uses the RADIUS protocol to request that the RADIUS server authenticate the user. If the RADIUS server approves the user's credentials, the firewall or SSL proxy allows the user access to the file server. Similar applications of RADIUS will invariably be used to support VoIP systems. User-based access controls can be enforced following RADIUS authentication to control URIs VoIP users may place calls to, or to control which users are allowed to place calls that incur toll charges.

Figure 8.3 illustrates a RADIUS exchange between a firewall and a RADIUS authentication server.

In Figure 8.3, we also illustrate the role RADIUS accounting [9] plays, and how it is performed. A RADIUS client can request that the RADIUS server begin accounting at any time, but will typically do so at the beginning of some user activity for which accounting or auditing is important. In the figure,

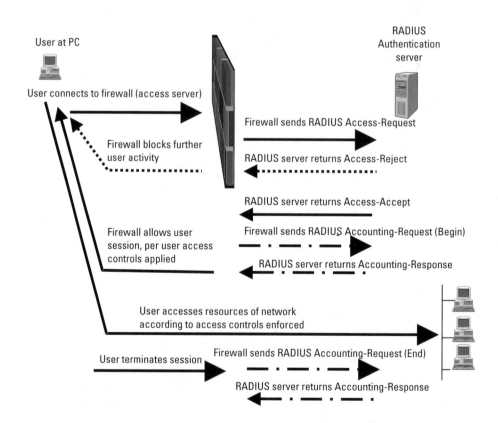

Figure 8.3.   RADIUS exchange.

accounting begins at the onset of a user session and continues for the duration of the session. Traffic utilization and time are common parameters, but the RADIUS Accounting RFC indicates that any attribute valid in a RADIUS Access-Request or Access-Accept packet is valid in a RADIUS Accounting-Request packet, including vendor-specific attributes (the only excepted attributes are those that could be used to compromise user accounts, such as a password or other credential). This flexibility allows administrators to use RADIUS accounting messages for the broader application of auditing (see Chapter 5) and provides opportunities for vendors to be creative when implementing accounting services for VoIP operators. Requirements for SIP accounting are considered in [10].

RADIUS has few built-in security services. RADIUS clients and servers authenticate by exchanging a message digest (unsigned) of a preshared secret. The RADIUS protocol does not have any confidentiality or integrity measures. RADIUS messages are typically protected by assuring that RADIUS exchanges are transmitted over physically isolated LAN (VLAN) segments behind firewalls or inside an authenticated and encrypted tunnel. RADIUS servers are considered part of the critical infrastructure and merit the kind of security measures identified for such servers in Chapter 7.

8.4 Conclusions

Authentication is an essential security service for application and network access, and system administration. Authentication is increasingly playing an important enabling role in admission control as well. Authentication can be performed at various and multiple levels of the Internet architecture, between users and applications, client endpoint devices and servers, and by peer servers.

The two authentication methods and protocols described in this chapter illustrate how users and endpoints are authenticated. In later chapters, we explain how these methods can be applied to VoIP systems and networks.

References

[1]   IEEE 802.1x Port-Based Access Control, www.ieee802.org/1/pages/802.1x.html.

[2]   Aboba, B., L. Blunk, J. Vollbrecht, J. Carlson, and H. Levkowetz, "Extensible Authentication Protocol (EAP)," RFC 3748, June 2004.

[3]   Aboba, B., and D. Simon, "PPP EAP TLS Authentication Protocol," RFC 2716, October 1999.

[4]   Funk, P., and S. Blake-Wilson, "EAP Tunneled TLS Authentication Protocol Version 1 (EAP-TTLSv1)," Internet-Draft, work in progress, February 2005.

[5] Palekar, J., et. al., "Protected EAP (PEAP) Version 2.0," http://josefsson.org/draft-josefsson-pppext-eap-tls-eap-10.txt.

[6] Cisco Systems, "Lightweight Extensible Authentication Protocol," http://www.cisco.com/application/pdf/en/us/guest/products/ps430/c1167/cdccont_0900aecd801764f1.pdf.

[7] Cisco Systems, "Extensible Authentication Protocol-Flexible Authentication via Secure Tunneling," http://www.cisco.com/application/pdf/en/us/guest/products/ps430/c1167/ccmigration_09186a00802030dc.pdf.

[8] Rigney, C., S. Willens, A. Rubens, and W. Simpson, "Remote Authentication Dial In User Service (RADIUS)," RFC 2865, June 2000.

[9] Rigney, C., "RADIUS Accounting," RFC 2866, June 2000.

[10] Loughney, J., and G. Camarillo, "Authentication, Authorization, and Accounting Requirements for the Session Initiation Protocol (SIP)," RFC 3702, February 2004.

9

Signaling Security

9.1 Introduction

Signaling in a VoIP network contains a great deal of sensitive information. Through signaling, users assert identities, and VoIP network operators authorize billable resources such as PSTN gateway and conferencing services. This chapter will discuss methods for securing the VoIP signaling path. We first discuss SIP signaling security and then H.323 signaling security.

SIP has a number of security mechanisms to protect signaling traffic from attacks against calling and called identities, including impersonation and eavesdropping for the purposes of calling pattern analysis, and path analysis. SIP uses several authentication mechanisms to prevent impersonation. SIP also recommends the use of encryption to prevent an attacker from monitoring calling patterns. Calling pattern monitoring is a form of traffic analysis. This is a highly sophisticated attack methodology. The attacker collects call signaling traffic and studies the calls placed to learn who places calls most frequently, to whom, and when. From the call data, the attacker tries to identify communicating parties, conversations, and calling windows that have some priority (e.g., convey time-sensitive information). One objective of such an attack is to identify a time where an attack can do most harm. For example, an attacker might learn enough to interfere with an auto-dialer program that notifies delivery truck drivers of pickup locations each day at 5:00 p.m.

For some business communications, it may be important to hide the path along which calls are placed, to prevent an attacker from identifying and focusing attacks on a critical point within a VoIP network. SIP also recommends encryption to keep call routing information (such as the `Via` fields in the SIP headers) confidential.

End-to-end encryption of SIP messages provides the best confidentiality and integrity characteristics, and SIP is flexible enough to accommodate multiple encryption schemes. However, in end-to-end encryption scenarios, only the called and calling parties can know the encrypting and signing keys. Therefore, it is not possible to encrypt all header information in SIP requests in responses end-to-end in all VoIP scenarios. For example, To and Via fields in SIP headers must be visible when calls are placed through SIP proxies. In such scenarios, cooperating proxies may employ hop-by-hop encryption.

9.2 SIP Signaling Security

SIP specifies a number of mechanisms to provide authentication, confidentiality and integrity security services. The first and simplest authentication mechanism described in the base SIP specification is HTTP digest authentication. As the name suggests, digest authentication uses a message digest or hash function to protect a shared secret as it is passed between the authenticator and the authenticating party during SIP session negotiation. SIP prescribes two algorithms for digest authentication: message digest 5 (MD5) and MD5 session mode (MD5-Sess). We look at these and compare them against basic authentication.

The original SIP RFC 2543 specifies pretty good privacy (PGP) as a means of securing SIP signaling. We discuss PGP briefly, and then discuss the use of secure MIME (S/MIME), TLS, and finally the use of the secure SIP URI scheme, SIPS.

9.2.1 Basic Authentication

RFC 2543 [1] allowed HTTP basic authentication, of the form:

```
WWW-Authenticate: Basic realm="mci.com"

Authorization: Basic QWxhZGRpbjpvcGVuIHNlc2FtZQ==
```

The response is simple base64 [2] encoding of the string username:password. Base64 is an encoding method that uses 6 bits per printable character or 65 possible symbols.

Basic authentication results in the cleartext transmission of user credentials, so the password can be easily recovered from the response string. Further, unless it is used with some form of timestamp, basic authentication offers no protection against reply attacks. If a UA retains support for Basic authentication, it could be used by a MitM attacker to obtain the user's password by generating a falsified Basic authentication challenge to the UA.

Basic authentication was deprecated in the successor specification for SIP, RFC 3261. It is possible to use Basic authentication securely in some scenarios; for example, parties can protect basic authentication using a secure transport or network layer tunnel such as TLS or IPSec. Endpoint authentication performed by transport of network layer tunnels can also be used where hop-by-hop authentication is desirable or necessary. However, even with a secure tunnel, credential challenges should use digest or stronger authentication, methods as described in the next sections, especially as digest authentication places such a minimal burden on servers and clients.

9.2.2 Digest Authentication

SIP borrows digest authentication from HTTP, where it is defined in RFC 2917 [3]. Digest authentication is used to authenticate user agents, and may be used to authenticate users to proxies, proxies to users, by use of the `Proxy-Authenticate`, `Proxy-Authorization`, `WWW-Authenticate`, and `Authorization` header fields.

We depict SIP digest authentication in Figure 9.1. In this example, one UA issues a SIP digest authentication challenge to another UA by sending a `401`

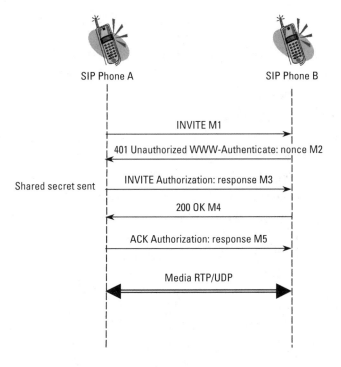

Figure 9.1 SIP digest authentication example between two user agents.

Unauthorized response. This response contains a WWW-Authenticate header field which contains the details of the digest challenge and also the nonce to be used for calculating the message digest. The UA sends an ACK to complete the SIP exchange. If the UA has credentials for the realm specified in the challenge, the request is resent with the credentials in an Authorization header field. Example WWW-Authenticate and Authentication header fields from the SIP specification are:

```
WWW-Authenticate: Digest realm="atlanta.com",
        domain="sip:boxesbybob.com", qop="auth",
        nonce="84a4cc6f3082121f32b42a2187831a9e",
        opaque="", stale=FALSE, algorithm=MD5

Authorization: Digest username="Alice",
        realm="atlanta.com",
        nonce="84a4cc6f3082121f32b42a2187831a9e",
        response="7587245234b3434cc3412213e5f113a5432"
```

The digest response is calculated by applying the hash function (by default, MD5) to the concatenation of the username, password, nonce, SIP method, and the request-URI. The complete set of parameters is shown in Table 9.1.

Table 9.1
Digest Challenge and Response Parameters

Parameter	Meaning
realm	Domain challenging
nonce	Random string provided in challenge used in response hashing algorithm
opaque	String generated by challenger and returned by client
stale	Flag used to indicate if nonce is stale
algorithm	Indicates whether MD5 or MD5-Sess is used
qop	Quality of protection, either auth for just authentication or auth-int for authentication and integrity
user	Username of user
uri	The request-URI
cnonce	Client-generated nonce, used in client authentication challenges to a proxy
nc	Nonce count
response	Message digest hash of shared secret
nextnonce	Nonce to be used in next authentication
rspauth	Response digest

As discussed in Chapter 2, MD5 [4] is a one way mathematical function that produces a fixed length digest of the message.

The exact MD5 algorithm is detailed in Section 9.2.2.1.

SIP digest authentication offers protection against replay attacks, in which an attacker captures a SIP request, and posing as a calling UA, resends this to a called UA. To protect itself against such attacks the challenger (called UA) must generate unique nonces and expire them as soon as they are used, or, alternatively, the challenger may use a timestamp. This appears to require that the challenger keep state for each authentication challenge, which could be used against the challenger by sending a flood of requests which would in turn generate a flood of challenges in the hopes of producing a memory overflow or buffer overflow. However, if a challenger uses a systematic method of generating nonces, a challenger can determine in a stateless way if a nonce is a valid one (one it generated) and whether it has expired. For example, a timestamp with some salt encrypted with a private key known only to the challenger could meet this requirement.

As with any challenge/response mechanism, limits should be placed on the number of failed authentication attempts sent by a particular user or from a particular IP address. If this limit is exceeded, additional requests should not be processed for a period of time (alternately, requests can be blocked indefinitely, i.e., until an administrator intervenes). A "block incoming requests" timer hampers brute-force attacks but does not mitigate them. The timer can be a fixed value, or it can be increased each time the failed attempts limit is exceeded.

If a responder provides a valid response but uses an invalid or out of date nonce, this can be communicated in the response using the `stale` flag along with a valid nonce.

This `401` challenge is also utilized by SIP registrar servers and for UA challenges. A proxy server can utilize this mechanism by using a `407 Proxy Authentication Required` challenge. Example header fields from the SIP specification are:

```
Proxy-Authenticate: Digest realm="atlanta.com",
      domain="sip:ss1.carrier.com", qop="auth",
      nonce="c60f3082ee1212b402a21831ae",
      opaque="", stale=FALSE, algorithm=MD5

Proxy-Authorization: Digest username="Alice",
      realm="atlanta.com",
      nonce="c60f3082ee1212b402a21831ae",
      response="245f23415f11432b3434341c022"
```

Using the `Authentication-Info` header field, it is possible to perform mutual authentication and integrity protection across the message bodies using digest. The `Authentication-Info` header field can be included in a 2xx response as shown in Figure 9.2. An example `Authentication-Info` header field is shown below:

```
Authentication-Info:
    nextnonce="47364c23432d2e131a5fb210812c",
    rspauth="23432d2e1b34343ee1212b402"
```

A nextnonce present in an `Authentication-Info` header field provides the nonce to be used in the next request, which can save another challenge and the associated messages. A `rspauth` contains an MD5 hash of the server's shared secret constructed as below, but leaving out the method. Figure 9.2 shows an example of mutual authentication using the `Authentication-Info` header field.

Any SIP method besides ACK and CANCEL can be challenged in this manner. An ACK can not be challenged because there is never a response to an ACK.

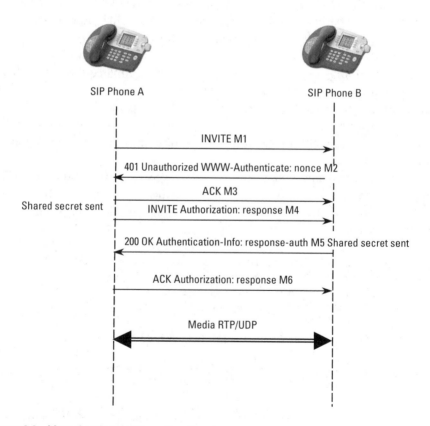

Figure 9.2 Mutual authentication using digest.

As a result, a UA should put the same credentials in an ACK as it did for the original INVITE, and proxies may not challenge the ACK. Although a response is issued to a CANCEL request, a CANCEL can not be resent since its CSeq count can not be incremented (the CSeq count in a CANCEL must match the request it is canceling). Also, the single hop nature of CANCEL makes challenges difficult. The only security that can be applied to a CANCEL is for a UA or proxy to try to determine if it came from the same source as the INVITE that it is canceling. As such, a secure system may decide not to accept CANCEL requests unless another authentication method is used.

A SIP UA may have a keyring with a number of credentials for different servers and services. The realm parameter is used to uniquely identity the credential, and represents a domain. When a digest challenge is received, a UA may use any credential matching the realm in the keyring. RFC 3261 defines an "anonymous" username with no password which may be tried. This allows a service to challenge all requests, but perhaps allow some limited service for anonymous users.

A UA should verify that the challenger is from the domain associated with the realm. For example, if a single hop TLS connection is used, the UA can validate the certificate of the proxy server or UA which is issuing the challenge.

Multiple digest challenges can be performed, for example, in a proxy chain. Note, however, that a UA cannot verify that the challenge comes from the appropriate domain if more than one domain is present.

The algorithm can have the values md5, md5-sess, or aka for MD5, MD5-session, or authentication with key agreement (AKA), which are described in the following sections.

9.2.2.1 MD5

When the qop is set to the value auth-int, digest authentication offers some integrity protection for the messages in either direction. Specifically, header elements and the message body used in the calculation of the WWW-Authenticate and Authorization header field response are protected against modification.

If qop=auth-int the response is the MD5 message digest of the following string:

MD5(username:realm:password):nonce:nc:cnonce:qop:
MD5 (method:uri:MD5(message-body))

This hash provides integrity over the message body, which is very useful when SDP data is present in a message body, for example.

If qop=auth the response is the MD5 message digest of the following string:

MD5(username:realm:password):nonce:nc:cnonce:qop:MD5(method:uri)

If qop is not present, the response is the MD5 message digest of the following string:

MD5(username:realm:password):nonce:MD5(method:uri)

Note that it is possible (and in practice, common) to store only the MD5 hash of the username, password, realm, and password on the authentication server instead of the actual password itself.

9.2.2.2 MD5-Sess

In MD5 session mode (MD5-Sess), a session key is calculated during the first challenge/response exchange and reused in all future challenge/responses:

A1 = MD5(username:realm:password):nonce:cnonce

This value of A1 is calculated once, then used as a session key for the rest of the session.

If qop=auth-int the response is the MD5 message digest of the following string:

A1:MD5(method:uri:MD5(message-body))

This hash provides integrity over the message body, which is very useful when SDP data is present in a message body, for example.

If qop=auth or if qop is not present, the response is the MD5 message digest of the following string:

A1:MD5(method:uri)

Authorization with key agreement (AKA) is a challenge/response mechanism used with a symmetric key. It is used in universal mobile telecommunications system (UMTS) networks. Its usage with HTTP digest is described in RFC 3310 [5].

9.2.3   Pretty Good Privacy

The original SIP specification RFC 2543 [1] defined a way to use pretty good privacy or PGP [6] for SIP privacy, authentication, and integrity. PGP is an encryption scheme invented by Phil Zimmermann for e-mail message security [7]. Its use has been extended to general purpose desktop encryption and signing. Commercial and freeware applications provide toolkits for signing and encrypting files and archives, and for secure deletion.

The original specification included the use of PGP in `WWW-Authenti-`
`cation, Authentication` header fields. A client (UA) signs digest authentication requests with his private key, and the recipient (for instance, a call server)
verifies the UA is who he claims to be by decrypting the request with the user's
public key. In addition, using the `Encryption` and `Response-Key` header
fields, PGP can be used to generate encrypted SIP message bodies between SIP
clients and servers.

The use of PGP has been deprecated in the latest specification in favor of
S/MIME, described in the next section. Few SIP implementations surfaced with
PGP, partly because the requirement to "canonicalize" or standardize the format
of requests over which the PGP signature was computed proved cumbersome,
and partly because the same "public key" administration that has hindered widespread deployment of PGP secured email hinders SIP.

9.2.4 S/MIME

S/MIME (Secure/Multipurpose Internet Mail Extensions) [8] provides a mechanism for providing integrity and confidentiality in SIP messaging. A UA using
S/MIME can digitally sign all or part of the message, and the digital signature
provides the recipient with the ability to determine if the message has been modified in transit. A message body can be S/MIME-encrypted so that it is not visible along the signaling path or to intermediate systems.

SIP UAs using S/MIME must support RSA as a digital signature algorithm, SHA-1 as a message digest hash, and AES (advanced encryption suite)
with 128 bit keys as a message (bulk) encryption algorithm [9]. RFC 3261
specified the 3DES encryption algorithm, but this has been obsoleted by RFC
3853, which requires AES [10] instead. Also note that RFC 3853 updates the
`Content-Transfer-Encoding` from `base64` to `8bit`.

The SIP specification does not discuss how users and organizations acquire
certificates. While certificates issued by trusted third parties are preferred, SIP
permits the use of self-signed certificates (see Chapter 12).

S/MIME can be used in the following ways in SIP:

- Encryption of a message body. SDP or other message bodies can be
  entirely encrypted. This might be used, for example, when keying material is carried in the SDP using the SDP Security Session Descriptions,
  described in Chapter 10.

- Privacy and integrity of the entire SIP message. The entire SIP message
  or response including the message body can be encrypted and carried as
  a message body. When received by the other party, the body can be
  decrypted and compared to the received SIP message.

For example, if Alice wants to send a message body in a SIP request that only Bob can see, Alice would encrypt it using Bob's public key. Instead of the normal message body, the SIP message would have:

```
Content-Type: applicaton/pkcs7-mime
 ;smime-type=envelope-data; name=smime.p7m

Content-Disposition: attachment; filename=smime.p7m
 ;handling=required
```

If Alice wants to sign a SIP message for authorization and integrity, it would generate and sign the entire SIP message including the message body using Alice's private key. The resulting SIP request would have:

```
Content-Type: multipart/signed
 ;protocol="applicaton/pkcs7-signature"
 ;micalg=sha1;boundary=bound34
```

Note that the boundary string bound34 could be any string. The first body would be of Content-Type:message/sip and the second body would be:

```
Content-Type: application/pkcs7-signature; name=smime.p7s
Content-Transfere-Encoding: 8bit
Content-Disposition: attachment; filename=smime.p7s
 ;handling=required
```

If Alice wants to send Bob a SIP message with some private SIP header fields only available to Bob and with a signature, Alice would encrypt a message/sip body with Bob's public key and sign the message using Alice's private key. The resulting message would have:

```
Content-Type: multipart/signed
 ;protocol="applicaton/pkcs7-signature"
 ;micalg=sha1;boundary=bound34
```

Again, note that the boundary string bound34 could be any string. The first body would be:

```
Content-Type: applicaton/pkcs7-mime
 ;smime-type=envelope-data
 ;name=smime.p7m
Content-Transfer-Encoding: 8bit
Content-Disposition: attachment; filename=smime.p7m
 ;handling=required
```

And the second body:

```
Content-Type: application/pkcs7-signature; name=smime.p7s
Content-Transfer-Encoding: 8bit
Content-Disposition: attachment; filename=smime.p7s
 ;handling=required
```

S/MIME is typically implemented using a toolkit as the S/MIME specification is quite complex and long. This approach is reasonable for PC and laptop soft clients, but does not seem applicable for getting S/MIME support into embedded devices and small footprint devices.

9.2.5 Transport Layer Security

SIP can utilize TLS [11] on a hop-by-hop basis to secure signaling messages from UAs to proxy and call servers. Consider a SIP session established between two UAs with two proxy servers in between. The SIP messages will go over three hops—UA to proxy, proxy to proxy, then proxy to UA. Each of these hops is a separate TLS session. Since the negotiation of a secure TLS tunnel is performed independently between the endpoints of the hop, each UA and proxy endpoint has its own set of credentials. Different TLS credentials can be used by UAs and proxies; for example, UAs could use any SIP message authentication to authenticate to proxies, but in theory, the proxies could authenticate each other using mutual authentication based on digital (RSA) certificates. Each tunnel in theory can also use different TLS cipher sets to choose different hash and encryption algorithms and key strengths (this is possibly useful in situations where export restrictions are a factor in encryption selection). The resulting end-to-end SIP messaging exchange utilizes TLS for confidentiality and integrity over each hop in the signaling path, but the contents of the SIP messages will be visible to both proxy servers.

SIP utilizes DNS NAPTR [12] records for selecting transport on successive hops. When a SIP URI such as `sip:alice@example.com` is specified, a DNS NAPTR lookup is performed to see if the example.com domain supports TLS transport. If so, the signaling request is submitted, and will have the value `TLS` specified in the `Via` header field.

The TLS handshake proceeds exactly as it does for secure web transactions. The server (proxy) offers its digital certificate and the UA attempts to verify the server certificate as follows:

1. Verify that the certificate has not expired;
2. Verify that the issuing CA is one the UA trusts;
3. Verify the `subjectAltName` covers the remote address:

a. If this is a request sent by the UA, the `subjectAltName` should match the address which the request was sent to (the server name resolved by SRV)

b. If this is a response received by the UA, the `subjectAltName` should match the address in the top `Via` header field

4. Verify that the IP address of the other side matches one resolvable using DNS.

TLS SIP requests use the default port 5061. A UA or a proxy examines the `Via` header field to determine whether TLS transport has been used on previous hops. If the application attempts to use TLS and the setup fails, the application is aware of this.

TLS uses TCP for transport. If any stream based transport is used with SIP, the `Content-Length` header field is mandatory, as it is the only delimiter between SIP messages in the stream.

If a TCP connection is opened to send a request, the response is sent by opening a new TCP connection to the port listed in the `Via` header field in the request instead of reusing the TCP connection already open to send the request.

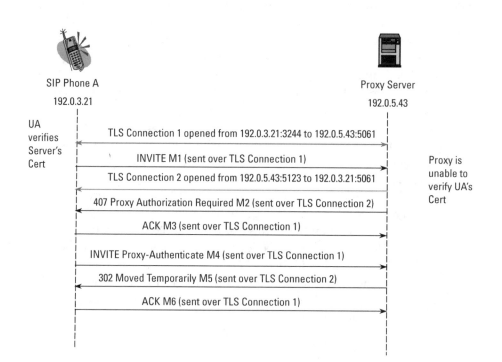

SIP Phone A
192.0.3.21

Proxy Server
192.0.5.43

UA verifies Server's Cert

TLS Connection 1 opened from 192.0.3.21:3244 to 192.0.5.43:5061

INVITE M1 (sent over TLS Connection 1)

TLS Connection 2 opened from 192.0.5.43:5123 to 192.0.3.21:5061

407 Proxy Authorization Required M2 (sent over TLS Connection 2)

ACK M3 (sent over TLS Connection 1)

INVITE Proxy-Authenticate M4 (sent over TLS Connection 1)

302 Moved Temporarily M5 (sent over TLS Connection 2)

ACK M6 (sent over TLS Connection 1)

Proxy is unable to verify UA's Cert

Figure 9.3 TLS connections without reuse.

This is shown in Figure 9.3. Because TCP uses two unidirectional flows rather than a single bidirectional connection, TCP connections are generally not reused for reverse routing at the application layer in SIP. For TLS connections, this is both inefficient and also introduces potential attacks in establishing a new secured connection when a secured one is open. For example, a TLS connection opened by a UA to a proxy server allows mutual authentication if the server provides a certificate and the UA provides a shared secret after a challenge. For responses, if a new TLS connection is opened, the client does not have a certificate to present, and the server has no way of challenging a response, so this connection is inherently less secure than the connection opened the other way.

An extension mechanism for connection reuse [13] has been defined to overcome this problem. An initiator of a connection includes a header parameter `alias` in the `Via` in the first request sent over the connection which requests the other side to establish a transport layer alias for reuse. In this manner, any requests or responses to be routed to the IP address and port number in the `Via` header will instead use the transport alias. This is shown in Figure 9.4.

Some call flows and implementation details about the use of TLS with SIP are in [14]. An example user certificate used for testing from [14] is shown in Figure 9.5.

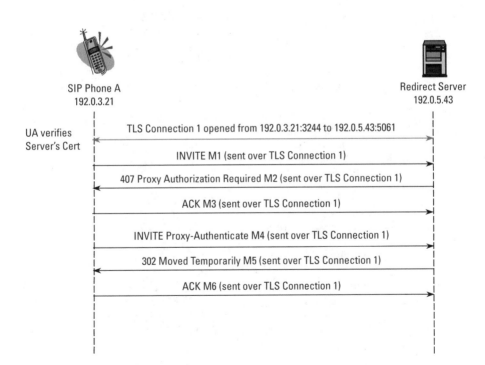

Figure 9.4 TLS connection reuse.

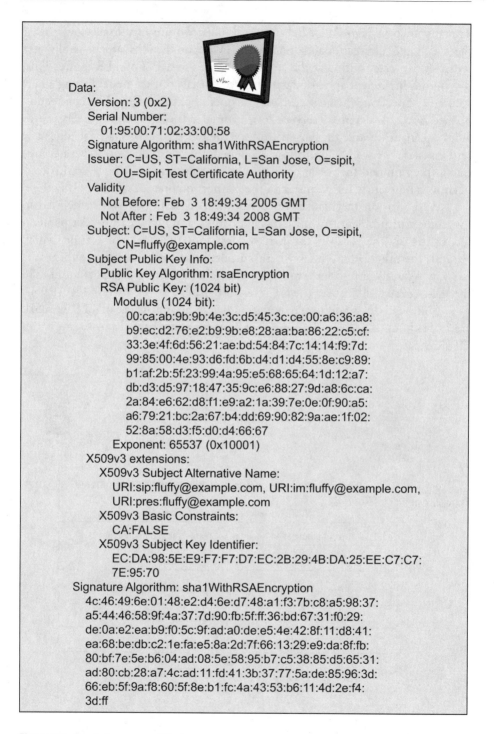

Data:
    Version: 3 (0x2)
    Serial Number:
        01:95:00:71:02:33:00:58
    Signature Algorithm: sha1WithRSAEncryption
    Issuer: C=US, ST=California, L=San Jose, O=sipit,
        OU=Sipit Test Certificate Authority
    Validity
        Not Before: Feb 3 18:49:34 2005 GMT
        Not After : Feb 3 18:49:34 2008 GMT
    Subject: C=US, ST=California, L=San Jose, O=sipit,
        CN=fluffy@example.com
    Subject Public Key Info:
        Public Key Algorithm: rsaEncryption
        RSA Public Key: (1024 bit)
            Modulus (1024 bit):
                00:ca:ab:9b:9b:4e:3c:d5:45:3c:ce:00:a6:36:a8:
                b9:ec:d2:76:e2:b9:9b:e8:28:aa:ba:86:22:c5:cf:
                33:3e:4f:6d:56:21:ae:bd:54:84:7c:14:14:f9:7d:
                99:85:00:4e:93:d6:fd:6b:d4:d1:d4:55:8e:c9:89:
                b1:af:2b:5f:23:99:4a:95:e5:68:65:64:1d:12:a7:
                db:d3:d5:97:18:47:35:9c:e6:88:27:9d:a8:6c:ca:
                2a:84:e6:62:d8:f1:e9:a2:1a:39:7e:0e:0f:90:a5:
                a6:79:21:bc:2a:67:b4:dd:69:90:82:9a:ae:1f:02:
                52:8a:58:d3:f5:d0:d4:66:67
            Exponent: 65537 (0x10001)
    X509v3 extensions:
        X509v3 Subject Alternative Name:
            URI:sip:fluffy@example.com, URI:im:fluffy@example.com,
            URI:pres:fluffy@example.com
        X509v3 Basic Constraints:
        CA:FALSE
        X509v3 Subject Key Identifier:
            EC:DA:98:5E:E9:F7:F7:D7:EC:2B:29:4B:DA:25:EE:C7:C7:
            7E:95:70
    Signature Algorithm: sha1WithRSAEncryption
        4c:46:49:6e:01:48:e2:d4:6e:d7:48:a1:f3:7b:c8:a5:98:37:
        a5:44:46:58:9f:4a:37:7d:90:fb:5f:ff:36:bd:67:31:f0:29:
        de:0a:e2:ea:b9:f0:5c:9f:ad:a0:de:e5:4e:42:8f:11:d8:41:
        ea:68:be:db:c2:1e:fa:e5:8a:2d:7f:66:13:29:e9:da:8f:fb:
        80:bf:7e:5e:b6:04:ad:08:5e:58:95:b7:c5:38:85:d5:65:31:
        ad:80:cb:28:a7:4c:ad:11:fd:41:3b:37:77:5a:de:85:96:3d:
        66:eb:5f:9a:f8:60:5f:8e:b1:fc:4a:43:53:b6:11:4d:2e:f4:
        3d:ff

Figure 9.5 Example user certificate.

Note that the `subjectAltName` contains three different identities, one with a SIP URI, another with a IM URI (for instant messaging) and another with a PRES URI (for presence). The `im` and `pres` URI schemes are defined in [15].

9.2.6   Secure SIP

Secure SIP [16] is a URI scheme `sips` that is used to request that TLS be used to secure every hop in the signaling path, from the caller to the *domain* of the called party (callee), although not necessarily all the way to the called party UA. An exception is made for the last proxy to UA hop that may be secured using some other means besides TLS. An example SIPS URI follows:

```
sips:c.babbage@calculatingmachine.org
```

Note that this is not equivalent to:

```
sip:c.babbage@calculatingmachine.org;transport=tls
```

While the latter is allowed, its use is discouraged (it is deprecated in IETF language) in favor of the SIPS URI scheme. Because of the multihop nature of SIP and the use of DNS to select transports by proxy servers, the second URI only provides a guarantee of a single hop of TLS; beyond that hop, the request could be routed over UDP, for example.

While end-to-end TLS is strongly recommended, RFC 3261 makes an exception for the last Proxy to UA hop within a domain to be secured using some other equivalent single hop transport. For example, a mobile phone's use of transport layer encryption or a Wireless LAN that requires clients to use WPA could meet this requirement.

A SIP proxy must support the `TLS_RSA_WITH_AES_128_CBC_SHA` cipher suite, which is AES cipher block chaining (CBC) mode with 128-bit encryption with SHA-1. Additional cipher suites can be supported. The SIP specification encourages use of mutual TLS authentication. When the SIPS URI scheme is used, either TCP or SCTP may be used, but not UDP. If a UA uses a Secure SIP URI, every hop must have confidentiality and integrity protection.

Following is an example of a complete secure SIP request:

```
INVITE sips:bob@192.0.2.4 SIP/2.0
Via: SIP/2.0/TLS server10.biloxi.com:5061;branch=z9hb4b43
Via: SIP/2.0/TLS
bigbox3.atlanta.com:5061;branch=z9hG4bK773.1
```

```
;received=192.0.2.2
Via: SIP/2.0/TLS pc33.atlanta.com:5061;branch=z9hG4bKn8
;received=192.0.2.1
Max-Forwards: 68
To: Bob <sips:bob@biloxi.com>
From: Alice <sips:alice@atlanta.com>;tag=1928301774
Call-ID: a84b4c76e66710
CSeq: 314159 INVITE
Contact: <sips:alice@pc33.atlanta.com>
Content-Type: application/sdp
Content-Length: ...
```

Note the presence of the TLS transport token in the Via header fields and the secure SIP URIs in the request-URI and From, To, and Contact header fields.

Secure SIP represents the optimal signaling security with SIP and will be used as the reference example throughout the rest of this book.

DTLS [17] has recently been standardized by the IETF and may be used to secure SIP signaling when UDP is used. Note, however, that Via header field tokens and SRV records for this new transport type have not yet been standardized.

9.3   H.323 Signaling Security with H.235

The security aspects of H.323 are described in H.235 [18], which will be summarized here. H.235 defines security for the following parts of H.323:

- H.225 RAS signaling between an endpoint and a gatekeeper;

- H.225 call signaling;

- H.245 control signaling.

H.235 also defines authentication and encryption of the media stream using a profile of SRTP, which will be discussed in Chapter 10.

As described in Chapter 3, an endpoint can locate a gatekeeper using the RAS GRQ request sent over multicast. The resulting GRQ / GCF exchange can perform a Diffie-Hellman exchange to generate a shared secret. This shared secret is then used for all future RAS exchanges with between the gatekeeper and the endpoint (RRQ, ARQ, etc.). This is shown in Figure 9.6.

Alternatively, if there is a shared secret between an endpoint and a gatekeeper, this shared secret can be used to authenticate the RAS connection. Methods defined include:

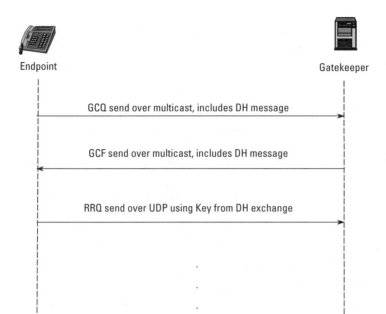

Figure 9.6 Establishment of H.225 RAS shared secret between endpoint and gatekeeper.

- Password with symmetric encryption;
- Password with hash;
- Certificate wth signature.

IPSec, TLS, or a proprietary transport can be used to secure the H.225 call signaling channel. When using TLS transport, H.323 uses port 1300 for the H.225 call signaling channel. Once this channel is established, a request for establishing a secure H.245 call control channel is then sent. Note that a request to establish a secure call control channel can be made over an unsecured call signaling channel.

References

[1]   Handley, M., H. Schulzrinne, E. Schooler, and J. Rosenberg, "SIP: Session Initiation Protocol," RFC 2543, March 1999.

[2]   Josefsson, S., "The Base16, Base32, and Base64 Data Encodings," RFC 3548, July 2003.

[3]   Franks, J., P. Hallam-Baker, J. Hostetler, S. Lawrence, P. Leach, A. Luotonen, and L. Stewart, "HTTP Authentication: Basic and Digest Access Authentication," RFC 2617, June 1999.

[4]   Rivest, R., "The MD5 Message-Digest Algorithm," RFC 1321, April 1992.

[5]   Neimi, A., J. Arkko, and V. Torvinen, "Hypertext Transfer Protocol (HTTP) Digest Authentication Using Authentication and Key Agreement (AKA)," September 2002.

[6]   Atkins, D., W. Stallings, and P. Zimmermann, "PGP Message Exchange Formats," RFC 1991, August 1996.

[7]   Elkins, M., "MIME Security with Pretty Good Privacy (PGP)," RFC 2015, October 1996.

[8]   Ramsdell, B., "Secure/Multipurpose Internet Mail Extensions (S/MIME) Version 3.1 Message Specification," RFC 3851, July 2004.

[9]   Peterson, J., "S/MIME Advanced Encryption Standard (AES) Requirement for the Session Initiation Protocol (SIP)," RFC 3853, July 2004.

[10]   Chown, P., "Advanced Encryption Standard (AES) Ciphersuites for Transport Layer Security (TLS)," RFC 3268, June 2002.

[11]   Allen, C., and T. Dierks, "The TLS Protocol Version 1.0," RFC 2246, January 1999.

[12]   Mealling, M., and R. Daniel, "The Naming Authority Pointer (NAPTR) DNS Resource Record," RFC 2915, September 2000.

[13]   Mahy, R., "Connection Reuse in the Session Initiation Protocol (SIP)," IETF Internet-Draft, work in progress, July 2004.

[14]   Jennings, C., and K. Ono, "Example Call Flows Using SIP Security Mechanisms," IETF Internet-Draft, work in progress, July 2005.

[15]   Peterson, J., "Common Profile for Presence," RFC 3859, August 2004.

[16]   Rosenberg, J., H. Schulzrinne, G. Camarillo, A. Johnston, J. Peterson, R. Sparks, M. Handley, and E. Schooler, "SIP: Session Initiation Protocol," RFC 3261, June 2002.

[17]   Rescorla, E., and N. Modadugu, "Datagram Transport Layer Security," RFC, June 2004.

[18]   "Security and Encryption for H-Series (H.323 and Other H.245-based) Multimedia Terminals," ITU-T Recommendation H.235v3, 2003.

10

Media Security

10.1 Introduction

In Chapter 9, we concentrated on the security of VoIP signaling. The security of the signaling is important, but if the media session established through signaling is not secure, VoIP call data, or, voice conversations, are vulnerable to monitoring, replay, and manipulation, including injection and deletion. A secure VoIP protocol that combines signaling and media into a single flow provides media security at little cost beyond the overhead associated with processing encryption. Such protocols, however, often prove extremely complex to design and implement. By decoupling of the signaling and media with protocols such as SIP and RTP, the individual protocols may be greatly simplified, but end-to-end security measures for signaling and media must be designed separately. The challenge of securing media streams includes:

- Completing a real time exchange of keys, crypto suite, and parameters without clipping the start of the conversation;
- Performing encryption, decryption, and authentication without introducing significant media latency or extra bandwidth;
- Rekeying without interrupting or adding delay to the session.

Due to the difficulty of solving these problems, the media exchange in a VoIP system often provides a lower level of security than the signaling (at least, initially). In this chapter, we consider the use of secure RTP (SRTP) to achieve integrity protection and confidentiality in the media session.

163

In some cases, media keying material is exchanged over the signaling channel. Multimedia internet keying (MIKEY) and SDP session descriptions fall into this category. Another approach, discussed at the end of the chapter, does not utilize the signaling path but uses the RTP media path itself to perform the key agreement.

10.2   Secure RTP

As we discussed in Chapter 3, a VoIP media stream is transported using the Real-Time Transport Protocol (RTP) [1], which utilizes User Data Protocol (UDP). Profiles for the use of RTP are defined in the RTP/AVP (Audio Video Profiles) specification [2].

The RTP Control Protocol (RTCP) provides information and some control over RTP sessions. The control aspect of RTCP is mainly limited to multicast sessions. For point-to-point media sessions common to VoIP, RTCP is used to exchange call quality reports.

RTP uses a 12-byte header field that consists of the following fields:

- *Version (V).* This 2-bit field is set to 2, the current version of RTP.

- *Padding (P).* If this bit is set, there are padding octets added to the end of the packet to make the packet a fixed length (suitable for block ciphers).

- *Extension (X).* If this bit is set, there must be one additional extension following the header (giving a total header length of 16 octets). Extensions are defined by certain payload types.

- *CSRC count (CC).* This 4-bit field contains the number of content source identifiers (CSRC) are present following the header. This field is only used by mixers that take multiple RTP streams and output a single RTP stream. Mixers which support this field allow endpoints to identify the current speaker in a conerence call, for example.

- *Marker (M).* This single bit is used to indicate the start of a new frame in video, or the start of a talk-spurt in silence-suppressed speech.

- *Payload type (PT).* This 7-bit field defines the codec in use. The value of this field matches the profile number listed in the SDP.

- *Sequence number (SEQ).* This 16-bit field is incremented for each RTP packet sent and is used to detect missing/out of sequence packets.

- *Timestamp.* This 32-bit field indicates in relative terms the time when the payload was sampled. This field allows the receiver to remove jitter and to play back the packets at the right interval assuming sufficient buffering.

- *Synchronization source identifier (SSRCI)*. This 32-bit field identifies the sender of the RTP packet. At the start of a session, each participant chooses a SSRC number randomly. Should two participants choose the same number, they each choose again until each party is unique.

- *Contributing source identifier (CSRC) list*. There can be none or up to 15 instances of this 32-bit field in the header. The number is set by the CSRC count (CC) header field. This field is only present if the RTP packet is being sent by a mixer, which has received RTP packets from a number of sources and sends out combined packets. A nonmulticast conference bridge would utilize this header.

SIP can establish multiple simultaneous media sessions, as, for example, audio and video streams. Each RTP stream will be received on different UDP ports and will have unique SSRC identifiers. An RTP session is uniquely indexed by the combination of the UDP port number on which the packets are received and the synchronization source (SSRC) identifier of the sender. As such, a two-way conversation between two endpoints is actually two RTP sessions, one in each direction.

Secure RTP (SRTP) [3] is a profile extension to RTP that adds confidentiality, authentication, and integrity protection to RTP sessions. SRTP takes an RTP stream, and adds encryption and integrity protection before handing the media stream to UDP for transport. SRTP uses symmetric keys and ciphers for media stream encryption, but does not provide any key management or generation functionality. Key management and exchange must be performed out of band from SRTP. We discuss one possible extension to SRTP to address this deficiency later in this chapter.

SRTP assumes that the communicating parties have already used a key management protocol to exchange or derive a set of master cryptographic keys for the set of ciphers to be used to protect the media stream. SRTP defines how session keys are generated from master cryptographic keys and how session keys are utilized, or refreshed, during the lifetime of the media session.

SRTP uses the AES algorithm in counter mode (CTR, also described as AES-CM in the SRTP specification) for encryption utilizing 128 or 256 bit keys. AES-CTR is similar to AES-CBC in that it turns a block cipher like AES into a stream cipher. Unlike AES-CBC, AES-CTR does not require feedback. Instead, the cipher text is produced by exclusive ORing (XOR) an RTP key stream with the plain text. The key stream is an encryption of a counter, initialized with a specific initialization vector (IV) of configurable length. This algorithm allows blocks to be calculated in parallel. The result is then included in the SRTP packet and occupies the same number of bytes. The IV is generated using a 112-bit salting key, the SSRC, and the SRTP packet index number (the RTP packet sequence number, SEQ, plus the rollover counter. ROC). The inclusion of the SSRC in the IV allows the same key to be used for multiple RTP media sessions, as each will have a different SSRC. As such, a single master secret could be used for both directions and for multiple media streams. Alternatively, each media stream in each direction could use a different master secret.

When message authentication is used, a HMAC SHA-1 hash is performed over the packet and added to the end, making the SRTP packet slightly larger than the RTP packet. This can be an issue for implementations that are trying to minimize media bandwidth requirements. Since RTP packets are usually much smaller than the maximum packet size (MTU), fragmentation of SRTP should not be an issue.

SRTP can use a number of master keys at the same time. Either communicating party can indicate which master key is to be used to encrypt a datagram in individual SRTP packets using the optional master key indicator (MKI) parameter.

The derivation of session keys from a master key is shown in Figure 10.1. At least one key derivation must be performed to obtain initial session keys. Subsequent key derivations may be performed depending on the cryptographic context of the SRTP stream. A master key and master salt are submitted to functions defined within SRTP to generate the session encryption key, the session salt key, and the authentication key. The session encryption key is used to drive the AES encryption algorithm. The session salt key is used as input to the IV. The authentication key is used for the optional HMAC SHA-1 message authentication function.

Figure 10.2 shows the makeup of an SRTP packet. The RTP header is not encrypted by SRTP. The RTP payload is encrypted. The additional fields optional introduced by SRTP (unshaded) are the master key index (MKI) and authentication tag.

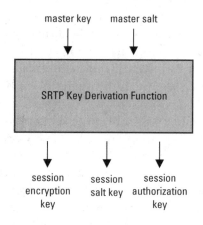

Figure 10.1 SRTP generation of session keys from master keys.

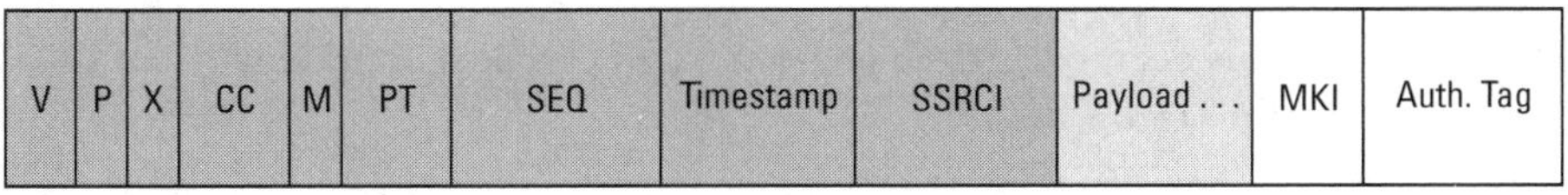

V	P	X	CC	M	PT	SEQ	Timestamp	SSRCI	Payload . . .	MKI	Auth. Tag

Figure 10.2 Secure Real-Time Protocol packet.

When present, the authentication tag provides integrity protection over the RTP header and payload (the shaded elements in the figure). Because one of the elements protected by the authentication tag is the RTP sequence number, authentication also provides replay protection. When authentication and encryption are applied, encryption is applied before authentication. The SRTP specification recommends a default of 80 bits for the authentication tag. However, the use of shorter and zero bit authentication tags is also discussed as a practical albeit less desirable implementation measure in applications where policy dictates that bandwidth preservation is to take precedence over strong authentication and in certain wireless applications that use fixed-width data links that are not capable of transferring the additional octet overhead of the authentication tag.

The MKI is a variable length field. Its length, value and use is defined by the key management service. Key management may use MKI for rekeying (refreshing keys). However, a typical point-to-point VoIP or even video over IP session will rarely need rekeying during the session. In situations where multiple Master keys are available, key management may use MKI to identify the master key that is to be used within the cryptographic context of a given SRTP stream. The MKI is not protected by the authentication tag.

SRTP is usable for extremely long-lived sessions, such as continuous broadcast video. As such, it defines the maximum lifetime of a master key to be 2^{48} SRTP packets. SRTP defines a mechanism for keeping track of SRTP packets beyond the 16-bit sequence (SEQ) number count defined and signaled in an RTP packet. An additional 32 bits known as the roll over counter (ROC) are defined and can be used to track the packet number. Besides the MKI, SRTP also defines a <from,to> mechanism for key lifetime.

SRTCP is also defined as a secure RTCP protocol. Although RTCP is typically only used to exchange quality reports in a point-to-point session, it is also used for multicast session control. As such, SRTCP message authentication is mandatory in the specification although not particularly useful in VoIP applications.

When message authentication is used, SRTP provides replay protection by keeping track of sequence numbers of received and authenticated SRTP packets. Using this replay list and, typically, a sliding window approach, SRTP determines whether an arriving packet is both authentic and "never before received." Conceptually, if the sequence number of an arriving packet matches an index in the replay list, it is deemed a replayed packet and can be discarded without causing any disruption to the session.

10.3 Media Encryption Keying

Here, we describe basic approaches to securing media streams. Some rely on the use of a preshared key (PSK) exchanged in advance of the session, while others use a public key infrastructure (PKI) and utilize public keys for encrypting key material. Another approach is to utilize a secured signaling channel to exchange keys or generate one-time session keys for media encryption and authentication.

All of the scenarios described in this section presume that the UAs have already agreed upon encryption ciphers and key lengths, and thus, concentrate on how keys are exchanged. We discuss how cryptographic context, cipher suites and configuration parameters can be exchanged securely at the start of a session in a subsequent section.

10.3.1 Preshared Keys

In the preshared keys approach, the UAs have previously exchanged secret keys for a symmetric cryptographic algorithm using an out of band method. For example, UAs within a group could be configured with a preshared key through security policy administration software, or as part of a group policy administration performed through active directory or a similar directory service. In some cases, the moderator of a VoIP conference call could distribute the shared secret key in a conference invitation. Since the keying material is not carried in the signaling, the signaling protocol does not need to be secured specifically to keep media encryption keys private. Also, since authentication is effectively provided by knowledge of the shared key, the signaling does not need to be authenticated to assure that media encryption keys are not altered. This encryption scheme is shown in Figure 10.3.

This approach is only useful within a small group, for a specific call, or for a single use of an application. For example, the moderator of a secure VoIP conference call may distribute a key to all participants for that single conference call.

In practice, preshared keys have the same characteristics as static passwords, and encrypting successive and multiple sessions using the same preshared key is poor design. For example, the longer a moderator uses the same preshared key, the greater the likelihood that the key may be unintentionally disclosed, "cracked" using a brute-force technique, or otherwise revealed. Once compromised, all media sessions encrypted with that key are vulnerable, including any previously placed calls that the attacker may have recorded. An alternative approach is to use preshared keys as part of the keying material, to generate new

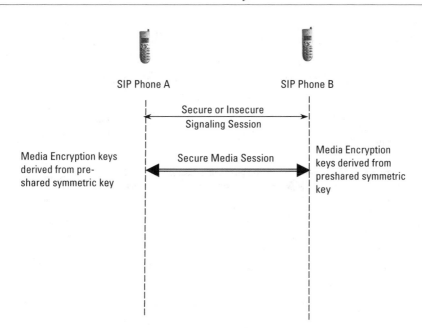

SIP Phone A   SIP Phone B

Secure or Insecure
Signaling Session

Media Encryption keys
derived from pre-
shared symmetric key

Secure Media Session

Media Encryption
keys derived from
preshared symmetric
key

Figure 10.3 Preshared key media encryption.

keys for each session, and to rekey long sessions based on key entropy. The alternative approach of key mixing makes decryption more difficult than if static keys were used, depending on how this is implemented. Consider 802.11 WPA-PSK[20]. If an attacker can capture the protocol exchanges used for session key derivation, he may be able to run a dictionary or brute force attack on the protocol to deduce the underlying preshared key. Once the attacker has that preshared key, he may be able to retrieve the session key and apply that key to decrypt (past or future) encrypted traffic.

10.3.2 Public Key Encryption

A simple public key model is shown in Figure 10.4. Alice and Bob utilize a public key infrastructure (PKI). Bob can use Alice's public keys to encrypt the RTP media packets. Alice has several options for making her public key available to Bob (and vice versa). She can encode her public key in the signaling path. Alice can publish her public key in a directory, typically as an element of a digital certificate issued by a trusted certificate authority, where Bob and others can retrieve it. In either case, by employing PKI, Bob and others can trust that the certificate (and hence the public key) is truly Alice's because they trust the certificate authority that issued Alice her certificate.

If public key encryption is used to encrypt successive and multiple media sessions, Alice must protect her private key from disclosure to prevent this

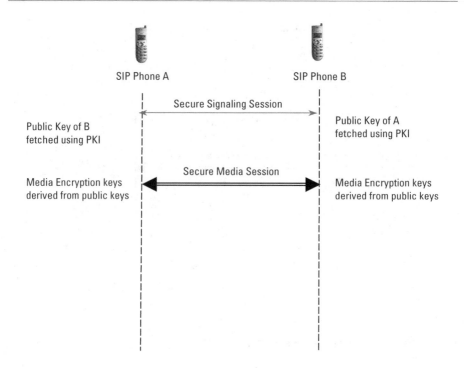

Figure 10.4 Public key media encryption.

solution. This is an improvement over preshared keys, but still not optimal. Moreover, public key encryption and decryption are more computationally intensive than symmetric key encryption and decryption for the same length keys. In practice, a better approach is for authenticated parties to securely negotiate a session key at the time of session setup and utilize the derived keys to encrypt the media for the duration of that session only.

We mentioned that encryption keys should be changed for each session. For long lasting, high bandwidth sessions, session keys should be changed at regular intervals (a practice known as rekeying) as well, so that an attacker breaking the ciphertext will only expose the session encryption key and not Alice's private (and hence, authentication) key. The resulting media exposure will also be limited to the part of the session that utilized the key, in short, a much lower exposure.

10.3.3 Authenticated Key Management and Exchange

An authenticated key management approach is shown in Figure 10.5. Alice and Bob authenticate themselves, and establish a secure signaling session. Over this secure channel, a set of session keys are exchanged or derived. These keys are

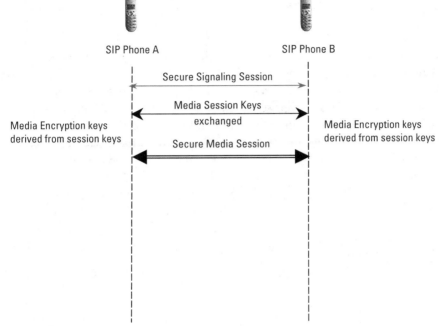

Figure 10.5 Secure media session establishment.

then used to generate media encryption keys which are used to encrypt and sign media packets.

Another option is to transport the keys over a secure SIP connection. As described in the previous chapter, secure SIP ensures that TLS will be used to cryptographically protect each hop in the signaling path. For example, keys for symmetric encryption of the media stream might be carried in an SDP message body in a SIP message secured by TLS. The original SDP specification [4] defined a k= attribute for the transport of a symmetric key. An example is shown below in which the encryption key is base64 encoded and carried in the SDP in the clear.

```
v=0
o=- 1313802769 1313803240 IN IP4 206.65.230.170
s=-
c=IN IP4 206.65.230.170
k=base64:9rtj2345kmfgoiew94kj34magposadfo23kdsfalaopqwot
t=0 0
m=audio 64028 RTP/AVP 100 0 8 18 98 101
a=fmtp:101 0-15
a=rtpmap:100 speex/16000
a=rtpmap:98 ilbc/8000
```

```
a=rtpmap:101 telephone-event/8000
a=sendrecv
```

A drawback to this approach is that the keying material is exposed to every proxy server in the exchange and is, as a result, susceptible to a MitM attack. Also, the only information conveyed in the k= field besides the key itself is the encoding used for the key. The set of SRTP options, such as the cipher suite, the key length, and whether both encryption and authentication are to be used, are not conveyed. Due to the limited extensibility of SDP, the k= attribute has now been deprecated and its use is not recommended in the updated SDP specification [5].

We discuss two newer approaches to key management in the following sections. In the first approach, extensions to SDP carry keys, cryptographic algorithms, and other parameters needed to configure the secure media session. In the second approach, SDP is used to carry a multimedia keying payload that securely carries the keying material.

10.4   Security Descriptions in SDP

Andreasen, Baugher, and Wing [6] describe a way to use the a=crypto attribute in SDP to carry SRTP keying and configuration information. Along with the keying material, the a=crypto attribute conveys the encryption and integrity protection algorithm, the master key lifetime, the master key index (MKI) number and the number of bits used to encode the MKI. An example is shown below:

```
a=crypto:1 AES_CM_128_HMAC_SHA1_80
inline:PS1uQCVeeCFCanVmcjkpPywjNWhcYD0mXXtextR|2^20|1:32
```

The first item in the attribute is a cipher set. Here, we choose AES counter mode as the encryption cipher, a 128 bit key length and SHA-1 80 bit as the HMAC authentication algorithm. In the next item, we concatenate the master key and master salt and encode the value using base64 following the inline: separator. The third item, separated from the keys by a | indicates that the master key is valid for 2^{20} SRTP packets, and final item also separated by a | specifies the the MKI (here, identified as 1, separated from the MKI length of 32 bits by a :. Since the MKI field is optional in SRTP, its presence, and length in bits must be signaled prior to the SRTP session.

We show a complete session description, including a security description, below:

```
v=0
o=jdoe 2890844526 2890842807 IN IP4 10.47.16.5
s=SDP Seminar
i=A Seminar on the session description protocol
u=http://www.example.com/seminars/sdp.pdf
e=j.doe@example.com (Jane Doe)
c=IN IP4 161.44.17.12/127
t=2873397496 2873404696
m=video 51372 RTP/SAVP 31
a=crypto:1 AES_CM_128_HMAC_SHA1_80
inline:d0RmdmcmVCspeEc3QGZiNWpVLFJhQX1cfHAwJSoj|2^20|1:32
m=audio 49170 RTP/SAVP 0
a=crypto:1 AES_CM_128_HMAC_SHA1_32
inline:NzB4d1BINUAvLEw6UzF3WSJ+PSdFcGdUJShpX1Zj|2^20|1:32
```

In this example, we propose two different master keys, one for audio and the other for video. The video stream uses an 80-bit message authentication code while the audio stream uses a 32-bit message authentication code. Note that the secure RTP session is signaled by the RTP/SAVP (for RTP secure audio video profile) token in each of the media (m=) lines, as defined in [3]. This approach for key exchange requires some other method to protect the keys, for example, S/MIME. If the destination is unable to decrypt the S/MIME, the INVITE will fail and the INVITE retried. (Failure modes are discussed in more detail in a later section.)

Note that if keying material is carried in SDP, we must provide integrity protection over the SDP or keying material could be altered or deleted by intermediaries.

10.5 Multimedia Internet Keying (MIKEY)

Multimedia Internet Keying (MIKEY)[7] is a key exchange protocol developed for the requirements of multimedia session security. There is a profile of MIKEY for SRTP and a usage and mapping to SDP is suitable for SIP and VoIP networks. MIKEY provides its own encryption and integrity protection, so it does not require that the entire SDP message body be encrypted. MIKEY supports a number of key exchange methods including preshared key, public key, and Diffie-Hellman key generation. The key exchange method chosen by the initiator must also be supported by the recipient, otherwise the exchange will fail.

MIKEY uses an offer/answer model and is transported using extensions to SDP defined in [8]. One party sends a MIKEY message to the other party during call setup, for example, in an INVITE message. The responding party answers with a MIKEY reply (e.g., in a 183 Session Progress or a 200 OK response). The exchange allows each UA to generate session keys and begin the encrypted SRTP media session.

MIKEY can be used in either of two ways. The first way is to negotiate separate security associations for each media stream, and the other is to negotiate a single security association for all media streams communicated over this common session. The relationship between SRTP, MIKEY, and RTP is shown in Figure 10.6.

MIKEY provides its own confidentiality, integrity and authentication services. However, MIKEY requires message authentication to assure that it remains associated with a specific SDP and SIP signaling message. Otherwise, an attacker could cut-and-paste valid MIKEY messages from other sessions, and possibly force key reuse.

Consider the SDP offer below, in which Alice offers both an audio and video media stream:

```
v=0
o= - 2890844526 2890844526 IN IP4 host.atl.example.com
s=
c=IN IP4 host.atl.example.com
```

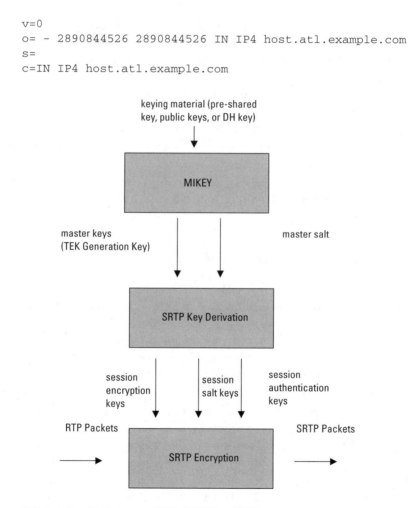

Figure 10.6 Relationship between SRTP, MIKEY, and RTP.

```
a=key-mgt:mikey Lkdlf3mdFLKES98fFk:wekDHJQodfje92dv...
t=0 0
m=audio 49170 RTP/SAVP 0 8 97
a=rtpmap:0 PCMU/8000
a=rtpmap:8 PCMA/8000
a=rtpmap:97 iLBC/8000
m=video 51372 RTP/SAVP 31 32
a=rtpmap:31 H261/90000
a=rtpmap:32 MPV/90000
```

The a=key-mgt SDP attribute extension is defined in [8]. The token mikey indicates that the key management protocol offered is MIKEY. The next element contains the base64 encoded MIKEY offer. If multiple key management protocols are supported, multiple key-mgt attributes can be listed (currently, only MIKEY is defined). Typically, a MIKEY exchange will result in the exchange or derivation of a single session key, used in each direction of the SRTP flow.

A secure RTP session is signaled by the RTP/SAVP (for RTP secure audio video profile) token in each of the media (m=) lines, as defined in [3].

Consider the following sample SDP answer:

```
v=0
o=bob 2808844564 2808844564 IN IP4 host.example.com
s=
c=IN IP4 host.example.com
a=key-mgt:mikey 4BjKdfkIjwekjfpo23GoTTe2#$56(fg...
t=0 0
m=audio 49174 RTP/SAVP 0
a=rtpmap:0 PCMU/8000
m=video 49170 RTP/SAVP 32
a=rtpmap:32 MPV/90000
```

In this answer, Bob accepts both the audio and video streams offers, and provides a MIKEY answer. If Alice accepts the answer, the secure RTP (SRTP) audio and video sessions begins. Note that in this example both media streams use the same master key. This is possible because each RTP session has a unique SSRC that results in a unique IV. If separate keys are to be used for each media stream, the a=key-mgt attribute line will be included after the media lines, as in this example:

```
v=0
o= - 2890844526 2890844526 IN IP4 host.atl.example.com
s=
c=IN IP4 host.atl.example.com
t=0 0
```

```
m=audio 49170 RTP/SAVP 0 8 97
a=key-mgt:mikey Lkdlf3mdFLKES98fFk:wekDHJQodfje92dv...
a=rtpmap:0 PCMU/8000
a=rtpmap:8 PCMA/8000
a=rtpmap:97 iLBC/8000
m=video 51372 RTP/SAVP 31 32
a=key-mgt:mikey ejhJheiLFekfkeifgj38doiFDk3wjfdg3ed...
a=rtpmap:31 H261/90000
a=rtpmap:32 MPV/90000
```

Two types of keys are described in MIKEY. They are traffic encryption keys (TEK) and TEK generation keys (TGK).

Traffic encryption keys are key inputs to the SRTP cryptographic application. When used in the context of SRTP, these keys are known as master keys. TEKs are generated using the TEK generation key (TGK) algorithm defined in the MIKEY specification. A successful MIKEY exchange results in the exchange or generation of one or more TGKs, which are then used by each UA to generate TEKs as input to SRTP.

A MIKEY message is created by first generating the MIKEY common header payload, one or possibly a series of payloads, and finally computing and appending a message authentication code (MAC)/signature to the message.

The common header payload (HDR) consists of the following fields:

- *Version* (the current version number of MIKEY is 1).

- *Data type* indicates the type of message. In the preshared key mode, this has a value of zero (0) for the MIKEY offer (initiator) message and one (1) for the MIKEY answer (verification), encoded as 7 bits. Other values are:

  - Initiator preshared key message;

  - Verification message of preshared key message;

  - Initiator's public key transport message;

  - Verification message of public key message;

  - Initiator's DH exchange message;

  - Responder's DH exchange message;

  - Error message.

- *Next payload* identifies the next payload. If there is no next payload, then 0 is used for last payload.

- *V flag* is a Boolean indicating whether a verification message is required.

- *PRF function* is a keyed pseudo random function used to generate encryption and authentication keys—MIKEY defines a default function.

- *CSB ID* (crypto session bundle identifier), a random identifier chosen by the initiator. A set of RTP and RTCP streams can share a crypto session bundle or can each have a separate one, depending on if the desire is to utilize the same keys and encryption algorithm. For the simplest case, all streams can use a single CSB.

- *#CS* indicates the number of crypto sessions within the CSB.

- *CS ID map type* (crypto session identifier map type) MIKEY defines the SRTP map type.

- *CS ID map info* (crypto session identifier map info) contains the SRTP SSRC (synchronization source identifier) and the rollover counter (ROC). Each media stream has an SSRC chosen by the initiator and used as the mapping between MIKEY and SRTP.

Requesting a response to a MIKEY initiator message is useful in that it provides a positive confirmation that the responder has received, processed, and understood the MIKEY message. It also provides an additional level of authentication. In one proposed mode of MIKEY known as RSA-R, the responder message actually contains the keying material.

The set of MIKEY payloads is shown in Table 10.1.

In addition to these payloads, the specification also defines a key data sub-payload, a key validity data payload, and a general extension payload.

10.5.1 Generation of MIKEY Message by Initiator

The initiator of the session generates the MIKEY offer. The first step of the key exchange is to use MIKEY to exchange master keys. The master keys will be used to generate session keys used to encrypt RTP packets. The MIKEY standard describes three ways of exchanging/generating traffic generation keys (TGKs):

1. Preshared keys;
2. Public keys;
3. Diffie-Hellman exchange.

While the preshared key approach requires the least computation, it does not meet the needs of a communication system in which secure sessions need to be established between large numbers of users who have no previous contact or interaction. The public key approach solves this problem, employing algorithms for Internet key exchange (IKE) [9]. For perfect forward secrecy (PFS), the Diffie-Hellman [10] approach generates unique keys each time the algorithm is

Table 10.1
MIKEY Payload Types

Payload	Description
HDR	Common header
KEMAC	Key data transport
PKE	Envelope data
DH	Diffie-Hellman
SIGN	Signature
T	Timestamp
ID	Identity
CERT	Certificate
CHASH	Certificate hash
V	Verification message
SP	Security policy
RAND	Random number
ERR	Error

invoked, eliminating the need for a long-term shared secret. (Self-signed certificates distributed by a SIP certificate server could also be used for this purpose and will be described in Chapter 12.)

10.5.1.1 Preshared Secret MIKEY Mode

Preshared Secret MIKEY Mode can be used in an initial authenticated key exchange if the UAs have already shared a secret key. This mode can also be used in a rekeying exchange once both parties have shared a secret key. For example, if an envelope key in the public key mode or a Diffie-Hellman derived key in the DH mode is available (see Sections 10.5.2.2 and 10.5.2.3), and then the preshared secret mode could be used for a rekeying exchange.

The payload sequence for generating an initiator's preshared secret MIKEY message is shown in Table 10.2.

The timestamp (T) is used by MIKEY to provide replay protection against an attempt by an attacker to cache a MIKEY message and inject it at a later time. To use a timestamp, each UA must maintain a reasonably accurate clock for this purpose. If NTP (Network Time Protocol) [11] or a local atomic clock source is not available, the time provided by a SIP registrar in a 200 OK response to a REGISTER message may be used. When using timestamp, SIP uses UTC (Universal coordinated time) to avoid time zone/local time complications.

Table 10.2

Payload Sequence for Preshared Secret Mode

Payload	Description
HDR	Header
T	Timestamp, encoded as NTP in UTC
RAND	Pseudo-random sequence used as a nonce, generated by the initiator
SP	Security policies are used to specify SRTP and set SRTP options, set by the initiator
KEMAC	Key data transport containing the encrypted traffic generation key (TGK) generated by the initiator

If the MIKEY initiator message requires a response, the response payload sequence will follow that shown in Table 10.3.

A key data transport payload, KEMAC, carries the encrypted TEK generation key (TGK). KEMAC includes a message authentication code for integrity protection, signed using an authentication key. The TGK is encrypted using an encryption key. The authentication and encryption keys are derived from a keyed pseudo-random function (known as the PRF and defined in the MIKEY specification) derived from the preshared key. The MAC is computed over the MIKEY Message (header, nonce, security policies, encrypted TGK) using a HMAC with SHA-1 [12, 13] and concatenated to the end of the message. This method of MIKEY initiator message generation is illustrated in Figure 10.7.

Note that the nonce is used as input to the PRF and also as input to the function which generates the TEK from the TGK.

10.5.1.2 Public Key Encryption MIKEY Mode

Public key encryption MIKEY mode does not use preshared symmetric keys. Instead, the MIKEY initiator generates a unique key called the envelope key. The initiator encrypts the envelope key using the public key of the MIKEY responder before transmitting the MIKEY message to the responder. The

Table 10.3

Response Payload Sequence for Preshared Secret Mode

Payload	Description
T	Timestamp
V	Verification message

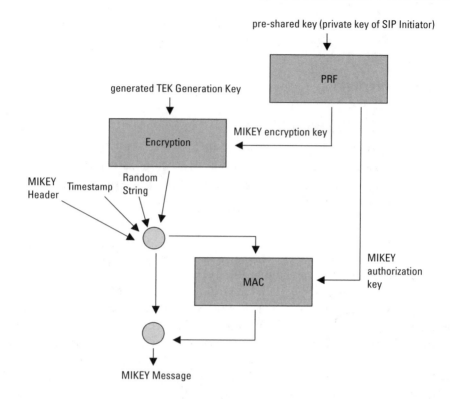

Figure 10.7 Preshared MIKEY generation sequence.

envelope key is used as keying material to generate an encryption key and an authentication key. The encryption key is used to encrypt the key data transport payload, KEMAC. The authentication key is used to authenticate the key data transport payload.

A hash computed over the entire MIKEY message is signed by the initiator using the initiator's private key to provide protection against MitM attacks. The signature algorithm is RSA PKCS#1 v1.5 [14] or RSA OAEP. The payload sequence for this public key mode is shown in Table 10.4.

In this mode, KEMAC carries the encrypted TEK generation key and Initiator's identity (SIP AOR URI). A message authentication code (MAC) is computed over KEMAC for integrity protection, and is signed using the authentication key derived from the envelope key. This is shown in Figure 10.8.

Public key MIKEY mode can be a problem to implement in VoIP systems where the public key of the responder may not be known before he answers. Consider a scenario where Alice calls Bob, but Bob's phone is forwarded to Carol's phone. Carol's phone may not be able to decrypt the MIKEY offer encrypted using Bob's public key. In fact, as will be discussed in Chapter 11, there is currently no agreed upon way for Carol to tell Alice that she has

Table 10.4
Payload Sequence for Public Key MIKEY Mode

Payload	Description
HDR	Header
T	Timestamp, encoded as NTP in UTC
RAND	Pseudo-random sequence used as a nonce
SP	Security policies that are used to specify SRTP and set SRTP options
KEMAC	Key data transport
PKE	Envelope data payload—envelope key encrypted using responder's public key
SIGNi	Signature of the hash computed over the entire MIKEY message using initiator's private key

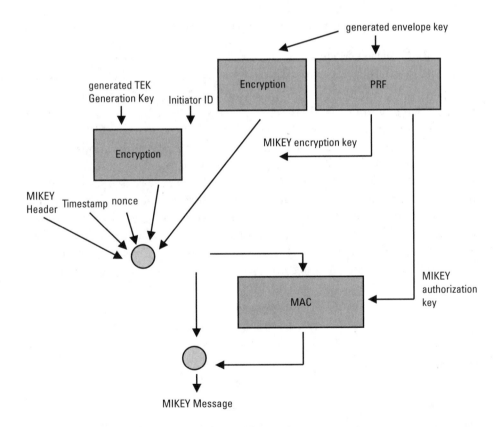

Figure 10.8 Public key MIKEY generation sequence.

answered instead of Bob. A work around for this is for the VoIP system to use call redirection instead of call forwarding. Bob's UA would then send a 302 Moved Temporarily response with Carol's URI in a Contact header field. Alice could then obtain Carol's certificate, calculate a new MIKEY offer, encrypt the new offer using Carol's public key, and then retry the request.

As an alternative, Ignjatic, Dondeti, Audet, and Lin [15] propose an additional mode which operates using public keys but does not require that the public key of the responder be known in advance. In this proposed MIKEY extension, known as RSA-R, the initiator doesn't generate keying material but requests that the responder generate the keying material instead. This mode may be useful in some multicast media scenarios, as well as some third party call control scenarios [16]. For example, consider that participants calling into a multicast media conference cannot propose the session key. Instead, the multicast bridge that generates the shared media stream can propose the session key.

The payload sequence is shown in Table 10.5. Two additional MIKEY data types are defined as initiator's public key message in RSA-R mode, and responder's public key message in RSA-R mode.

The setting of the verification bit V is required.

10.5.1.3 Diffie-Hellman Mode

Using the Diffie-Hellman MIKEY mode, UAs exchange information that enable each party to derive the same, unique session key. The Diffie-Hellman exchange has a particularly unique and attractive property. The session key is wholly derived using modular arithmetic. While both parties participate in the key generation process, the key itself is never transmitted over the communications medium by either party. Diffie-Hellman also has the desirable property of providing perfect forward secrecy (PFS). PFS is not guaranteed when pre-shared secret mode is used. If this preshared secret is compromised, all sessions established using this secret are compromised. PFS is also not guaranteed when publickey mode is used because again, if a private key is compromised, all keys exchanged using this key are compromised. However, if a key generated using Diffie-Hellman is compromised, only the session(s) using that key are compromised, Specifically, if an

Table 10.5
Payload Sequence for Public Key RSA-R MIKEY Mode

Payload	Description
HDR	Header
T	Timestamp, encoded as NTP in UTC
SIGNi	Signature of the hash of the entire MIKEY message using initiator's private key

attacker "cracks" a DH derived key he will be unable to decrypt previous messages that he may have captured and he will be unable to decrypt any future messages he might capture based on having cracked a single DH derived key, except through additional successful key attacks.

The payload sequence for this exchange is shown in Table 10.6.

The Diffie-Hellman group and value length used is carried in the DHi payload and can be OAKLEY 5 or optionally OAKLEY 1 or 2. The SIGNi signature is critical in that it protects against a MitM attack.

DH with HMAC authentication [17] provides an alternative protection against MitM attacks that does not require a public key infrastructure.

10.5.2 Responder Processing of a MIKEY Message

The initiator of the session generates the MIKEY offer. Depending on which mode is used, the processing will be slightly different, as discussed in the following sections.

10.5.2.1 Preshared Secret Mode

When preshared secret MIKEY mode is used, the receiving UA parses the MIKEY message and separates the header, timestamp, nonce, and MAC from the KEMAC. The receiving UA uses the timestamp to ensure that the message is not a replay attempt by an attacker. The receiving UA uses the shared secret key to generate the MIKEY encryption and authorization key, and next uses the authorization key to validate the integrity of the MIKEY message (check the MAC). If the message is authentic, the receiving UA uses the encryption key to extract the TGK to be used as the session key.

If required, the receiving UA generates a response and sends it to the initiator UA.

Table 10.6
Payload Sequence for Diffie-Hellman MIKEY Mode

Payload	Description
HDR	Header
T	Timestamp, encoded as NTP in UTC
RAND	Pseudo random sequence used as a nonce
SP	Security Policies which are used to specify SRTP and set SRTP options.
DHi	Diffie-Hellman value
SIGNi	Signature of the hash computed over the entire MIKEY message using Initiator's private key

10.5.2.2 Public Key Mode

When public key MIKEY mode is used, the receiving UA parses the MIKEY header and separates the timestamp, nonce, security policies, envelope data, and signature from the KEMAC. The receiving UA uses the timestamp to ensure that the message is not being replayed by an attacker. The receiving UA next uses the public key of the initiator to validate the signature. If the signature is valid, the receiving UA processes the authenticated MIKEY message. The receiving UA uses the responder's private key to decrypt the envelope data to obtain the envelope key, and next uses the envelope key to generate the encryption and authorization keys using the PRF function. The receiving UA uses the encryption key to decrypt the KEMAC and compares the SIP identity contained in the KEMAC to the initiator's SIP identity. If the two are the same, the receiving UA accepts the MIKEY message.

If required, the receiving UA generates a response and sends it to the initiator UA.

If the RSA-R mode is used, the payload sequence will be as shown in Table 10.7.

The header payload, HDR, in the responder message must use the same CSB as the initiator message, and will have CS fields filled in.

10.5.2.3 Diffie-Hellman Exchange Mode

When the Diffie-Hellman exchange mode is used, the responder generates a response that contains the responder's public exponent and the same set of payload elements as the initiator's DH message. Using the values contained in the two DH messages and local information (such as private keys), each UA calculates the same, unique session key (see Section 2.2.4.2).

Table 10.7
Payload Sequence for Public Key RSA-R MIKEY Mode

Payload	Description
HDR	Header
GenExt	General extension used in group key establishment
T	Timestamp, encoded as NTP in UTC
RAND	Pseudo-random sequence, used as a nonce
SP	Security policies are used to specify SRTP and set SRTP options.
KEMAC	Key data transport
PKE	Envelope data payload— envelope key encrypted using initiator's public key
SIGNr	Signature of the entire MIKEY message using responder's private key

SRTP policies are set using the MIKEY security policy data payload. These include:

- Encryption algorithm (NULL, AES-CM, AES-F8);
- Session encryption key length (depends on algorithm);
- Authentication algorithm (NULL or HMAC-SHA-1);
- Session authorization key length (depends on MAC);
- Session salt key length;
- SRTP pseudo-random function (AES-CM);
- Key derivation rate;
- SRTP encryption toggle on/off;
- SRTP encryption toggle on/off;
- Sender's FEC order (forward error correction);
- SRTP authentication toggle on/off;
- Authentication tag length (bytes);
- SRTP prefix length (bytes).

An example set of policies might be:

- Encryption algorithm: AES-CM;
- Session encryption key length: 128 bits;
- Authentication algorithm HMAC-SHA1;
- Session authorization key length (depends on MAC);
- Session salt key length: 112 bits;
- SRTP pseudo-random function: AES-CM;
- Key derivation rate: 2^{48};
- SRTP encryption on;
- Sender's FEC order: FEC then SRTP;
- SRTP authentication on;
- Authentication tag length: 32;
- SRTP prefix length: 0.

10.6   Failure and Fallback Scenarios

Most VoIP systems today do not support secure media sessions. As endpoints become more sophisticated, there will be an interim period where secure and insecure endpoints must interoperate. In some cases, an initiator UA will not know in advance whether or not a recipient UA supports secure media. Some VoIP implementations and service deployments will likely allow an initiator UA to accept an insecure session if the recipient UA is not capable of establishing a secure session. This backwards compatibility requirement must be deployed with care, since it creates an opportunity for a MitM attacker to force a downgrade from a secure session and monitor the media session. To protect against this threat, UAs should use an integrity mechanism, such as the `Identity` header field described in Chapter 11.

If the Security descriptions in SDP approach is used, the initiator UA must encrypt the entire SDP offer using S/MIME since the `a=crypto` attribute has the keying material in plain text. One approach includes a multipart MIME message body containing an encrypted SDP offer with keying material and an unencrypted insecure SDP offer. A drawback to this approach is that many UAs do not support multipart MIME and do not give the correct error response when they receive such a message. This is shown in Figure 10.9 below. Having received the `415 Unsupported Media Types`, A has no choice but to give up on the secure session or fall back to an unsecure session by sending a new `INVITE` with a single message body part with no S/MIME.

If the MIKEY approach is used, a secure and insecure media session can be offered in the same SDP offer. In this approach, the MIKEY offer is encrypted, and the rest of the SDP is transmitted in the clear. This does not require that both UAs support both S/MIME and multipart MIME, possibly leading to better interoperability.

Some other failure scenarios are listed below:

1. *The responder's certificate cannot be retrieved.* This could be caused by a failure to obtain the certificate from responder or the PKI infrastructure, or due to a failure with a SIP certificate server, as discussed in Chapter 11. In this case, SDP session descriptions cannot be used, since the SDP offer cannot be encrypted using S/MIME. If using MIKEY, only the public key RSA-R or Diffie-Hellman modes can be utilized.

2. *The responder does not support MIKEY.* This could be detected if all secure AVP (SAVP) media streams are declined. Note that a misbehaving UA might accept a SAVP offer, but treat it like a normal AVP media stream. In this case, no MIKEY response would be present.

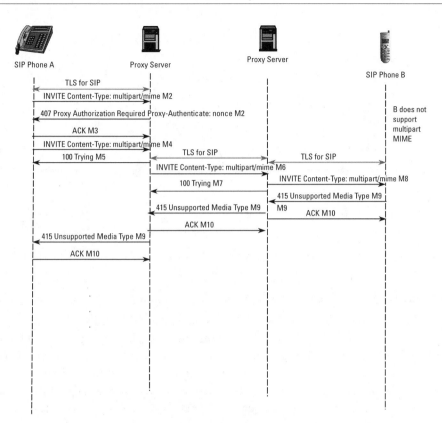

Figure 10.9 Failure and fallback to RTP.

SDP session descriptions could be used if the responder supported that method of key exchange.

3. *The responder does not support the MIKEY mode of the offerer.* All MIKEY modes except pre-shared key mode are optional. Currently, there are no mechanisms to discover via signaling which modes are supported before making the MIKEY offer. This looms as a major interoperability problem for MIKEY.

4. *The responder is unable to obtain the initiator's certificate or is unable to decrypt the MIKEY offer even with the certificate.* For example, if the crypto algorithm in the certificate is not supported by the responder. This would result in the declining of the SAVP media lines and a MIKEY error response generated, as listed below.

5. *The MIKEY offer is not acceptable.* As above, the SAVP media streams would be declined and a MIKEY error response generated. The set of Error payloads that can be returned in a MIKEY answer include the following:

- Authentication failure;
- Invalid timestamp;
- PRF function not supported;
- MAC algorithm not supported;
- Encryption algorithm not supported;
- Hash function not supported;
- DH group not supported;
- ID not supported;
- Certificate not supported;
- Security policy type not supported;
- Security policy parameters not supported;
- Data type not supported;
- Unspecified error condition.

6. *The MIKEY verification message is not acceptable.* MIKEY uses a strict offer/answer exchange that is limited to a single round-trip message exchange. As a result, there is no way to generate an error response to a MIKEY verification message. Instead, a re-INVITE might be sent with a new MIKEY initiator message to try for an acceptable mode of operation. Alternatively, a BYE could be sent to tear down the session.

7. *SRTP packets cannot be decrypted.* The UA might try another MIKEY exchange or send a BYE to tear down the session.

10.7   Alternative Key Management Protocol—ZRTP

All the key management mechanisms described thus far use the signaling channel as transport for the keying material. This is a very different approach from the one used by current PSTN secure telephones. Secure telephones establish a normal telephone call with another secure telephone, perform key management in-band in the audio channel to negotiate the security parameters, and then switch over to an encrypted session. This approach has been proposed and demonstrated for VoIP by PGP inventor Phil Zimmermann, in his *zfone* SIP user agent [18].

The zfone client is based on Zimmermann's PGPfone secure telephone developed in 1997. Zfone uses the RTP media session established between two zfone clients exclusively to perform key management and to negotiate encryption and authentication algorithms before switching to SRTP. The zfone implementation performs a Diffie-Hellman exchange using RTP header extensions.

This effectively makes SRTP a self-contained protocol that does not require an external key management protocol.

This key agreement scheme needs protection from an active DH MitM attack in which the attacker inserts himself in the middle of two zfone clients and performs separate DH agreements with both clients. Integrity protection using a signed hash can be used to verify that the DH agreement was performed with the other party and not an attacker. If reliance on a PKI infrastructure is a concern, other methods can be considered.

Secure PSTN telephones use a voice authentication digest. A series of digits are generated by each client based on a hash of the DH shared secret. Each party reads the digits to the other party. If either party reads incorrect digits, then the parties have not agreed on a common DH shared secret and they can conclude that a MitM attack is taking place. An attacker who attempts to generate the voice authentication digest will not be able to read out the digits in the voice of the party he attempts to impersonate. Note that this approach works even if the two parties have never spoken before. Each party simply needs to confirm that the voice which read out the voice authentication digest is the same voice used for the rest of the conversation. Of course, this approach is subject to the "Rich Little" attack in which an attacker could attempt to change their voice to match the desired voice. Zfone uses a similar voice authentication digest approach.

In addition to the protection provided by the voice authentication digest, the zfone protocol uses a long term shared secret cached from previous zfone sessions to generate the current shared secret. When the voice authentication digest is not used, this cached secret provides a similar level of security as SSH when used in the leap-of-faith mode: as long as there was no active MitM attacker in the initial session between the two parties, the cached secret provides protection against any future MitM attacks. Note that an active MitM attack involves the attacker actually participating in two DH key agreements at the start of the session, as passive eavesdropping alone does not result in a compromised session. However, if the cached secret is lost by either party, the protection goes away and a new shared secret must be generated. This provides the opportunity for a MitM attacker to make it appear that the shared secret has been lost, force a new session, and provide another opportunity to launch an attack.

This approach provides confidentiality for the media session, but not authentication. However, a shared secret generated in the signaling exchange could be mixed in with the DH secret to produce a key to provide confidentiality and authentication.

The choice of performing the key agreement in RTP instead of in the signaling is an interesting one. While there are many different signaling protocols for VoIP, including proprietary ones, almost all VoIP systems use RTP for the media. The zfone approach thus allows secure sessions to be established even when the endpoints do not share a common signaling protocol.

This protocol extension to SRTP may circumvent the anticipated interoperability problems associated with MIKEY. At the time of publication, the zfone approach is being considered for standardization within in the IETF, as ZRTP [19].

10.8 Future Work

The protocols and call flows described in this chapter are the newest and are the least verified by actual implementation. In particular, there are few implementations of MIKEY; the use of MIKEY with SIP, SDP, and SRTP is still under investigation; and the number of alternatives for MIKEY poses a problem for interoperability. Hopefully, future Internet-Drafts will piece these components together to create a manageable framework for initiating secure call flows. As always, new approaches, such as Phil Zimmermann's zfone, will surface to challenge early incumbents.

References

[1]   Schulzrinne, H., S. Casner, R. Frederick, and V. Jacobson, "RTP: A Transport Protocol for Real-Time Applications," IETF RFC 3550, July 2003.

[2]   Schulzrinne, H., and S. Casner, "RTP Profile for Audio and Video Conferences with Minimal Control," RFC 3551, July 2003.

[3]   Baugher, M., D. McGrew, M. Naslund, E. Carrara, and K. Norrman, "The Secure Real Time Transport Protocol", RFC 3711, March 2004.

[4]   Handley, M., and V. Jacobson, "SDP: Session Description Protocol," RFC 2327, April 1998.

[5]   Handley, M., V. Jacobson, and C. Perkins, "SDP: Session Description Protocol," IETF Internet-Draft, work in progress, February 2005.

[6]   Andreasen, F., M. Baugher, and D. Wing, "Session Description Protocol Security Descriptions for Media Streams," work in progress, February 2005.

[7]   Arkko, J., E. Carrara, F. Lindholm, M. Naslund, and K. Norrman, "MIKEY: Multimedia Internet KEYing," RFC 3830, August 2004.

[8]   Arkko, J., E. Carrara, F. Lindholm, M. Naslund, and K. Norrman, "Key Management Extensions for Session Description Protocol (SDP) and Real Time Streaming Protocol (RTSP)," Internet-Draft, work in progress, November 2004.

[9]   Hoffman, P., "Algorithms for Internet Key Exchange version 1 (IKEv1)," IETF RFC 4109, May 2005.

[10] Diffie, W., and M. E. Hellman, "New Directions in Cryptography," *IEEE Transactions on Information Theory*, Vol. IT-22, No. 6, June 1977, pp. 74–84.

[11] Mills, D., "Network Time Protocol (Version 3) Specification, Implementation and Analysis," IETF RFC 1305, March 1992.

[12] Krawczyk, H., M. Bellare, and R. Canetti, "HMAC: Keyed-Hashing for Message Authentication," RFC 2104, February 1997.

[13] NIST, FIPS PUB 180-1: Secure Hash Standard, April 1995.

[14] Kaliski, B., "PKCS #1: RSA Encryption Version 1.5," IETF RFC 2313, March 1998.

[15] Ignjatic, D., L. Dondeti, F. Audet, and P. Lin, "An Additional Mode of Key Distribution in MIKEY: MIKEY-RSA-R," IETF Internet-Draft, work in progress, July 2005.

[16] Rosenberg, J., J. Peterson, H. Schulzrinne, and G. Camarillo, "Best Current Practices for Third Party Call Control (3pcc) in the Session Initiation Protocol (SIP)," IETF RFC 3725.

[17] Euchner, M., "HMAC-authenticated Diffie-Hellman for MIKEY," IETF Internet-Draft, work in progress, April 2005.

[18] http://www.philzimmermann.com/zfone.

[19] Zimmermann, P., and A. Johnston, "ZRTP," IETF Internet-Draft, February 2006.

[20] Moskowitz, B., "Weakness in Passphrase Choice in WPA Interface," *WiFi Network News* (WNN), http://wifinetnews.com/archives/002452.html.

11

Identity

11.1 Introduction

The concepts of identity and identity verification are key to communications in general, and especially to secure communications. In Chapter 8, we discussed the role authentication plays in verifying identities. This chapter considers what authorization for identity means, how it is performed, and how it is asserted. The chapter begins with a discussion about names, and how a name is allocated. We discuss namespace management in the telephone network, how Internet names are assigned, and look briefly at H.323 names. We consider identity approaches for SIP VoIP systems and compare several alternatives. This chapter concludes by demonstrating how a SIP VoIP system can employ a cryptographically secure interdomain identity mechanism.

11.2 Names, Addresses, Numbers, and Communication

In Chapter 5, we introduced the concept of identity. A "name" is used to identify a user, service, or device. In order for a name to be used for communication, a name is typically resolved to an address. In the Internet, we use a specific type of name, a domain name, which is an ASCII text representation of one or more IP addresses. For example, before the web page associated with the name www.artechhouse.com can be retrieved by a browser, the name must be resolved using the domain name system (DNS) to an IP address. In the PSTN, telephone numbers tend to be treated as both names and addresses. This is changing with the integration of VoIP and traditional voice networks.

11.2.1 E.164 Telephone Numbers

Telephone numbers can be thought of as endpoint identifiers in the PSTN. Identity values (such as E.164 telephone numbers) are transported by various PSTN protocols such as integrated services digital network (ISDN) and ISDN user part (ISUP). ISDN is a UNI protocol often used by PBXes while ISUP is the signaling protocol part of SS7. The PSTN carries various kinds of identifiers including calling party, automatic number identification (ANI), and digital number information service (DNIS). End users of the PSTN are primarily familiar with the calling line identifier (CLID) also known as caller ID. This identity information is signaled over a two wire telephone loop between the second and third rings and can be decoded and displayed on telephones with a display. On cellular phones, the calling party is signaled using the wireless signaling protocol. Additional CLID services beyond numeric display include subscriber name display (which may be a business or individual but is always the billed party), incoming call blocking and call refusal notification. Some users use CLID logs instead of an answering machine, checking logs and using them to return missed calls.

The PSTN is effectively a single trust domain. At the edges of the network, telephone switches introduce CLID information, which is then transported end-to-end. Certain ISDN interfaces prove exceptions to this rule, as in the case where the end user can provide the calling station identity, which is then validated by the telephone switch. This deployment worked well when a single or only a few companies ran the telephone network. In many parts of the world, a single government entity runs the telephone network. In the United States, for most of last century, the Bell system ran the telephone network.

In today's deregulated PSTN with thousands of competing carriers, this trust model is harder to maintain. If a trusted network element introduces telephone calls with invalid or wrong CLID information, this information can be passed as trusted throughout the network. PSTN operators can only rely on transitive trust, and assume that all other PSTN operators are diligent in correctly configuring their subscriber interfaces and protecting their networks from impersonation attacks.

In the PSTN, a telephone number is effectively an address. It provides routing information used directly to find the terminating PSTN switch which provides telephone service for the called party. However, with local number portability (LNP), a telephone number is no longer the exclusive source of routing information in the PSTN. Once geographic number portability is introduced, for instance, the user's ability to keep a telephone number when moving to a different geographic location, telephone numbers will simply become names and will always require resolution before they can be used for routing.

Some telephone numbers may also not be unique. For example, an enterprise or campus can assign telephone extension numbers which only have significance within the enterprise or campus. Some VoIP providers issue numbers, but these numbers are not dialable on the telephone network.

11.2.2 Internet Names

RFC 822, standard for the format of ARPA Internet text messages [1], introduces the concept of computer user names within the context of electronic mail, of the form:

```
user@host.network
```

The `host.network` element is generally assumed to be either a fully qualified domain name (hhi.corecom.com) or a label (corecom) within a top-level domain (com).

Internet names use the hierarchical domain name service [2] to resolve domain names to IP addresses. There is a single, authoritative root for the Internet domain name system, referred to as dot or ".". Beneath this root, the Internet Corporation for Assigned Names and Numbers (ICANN) oversees a set of generic top-level domain (gTLD) labels, such as .com, .net, .org, .biz, .aero, and country code top-level domain (ccTLDs) labels, such as .uk, .us, .ca, .au. Second-level domain names are registered under the TLDs labels through registrars. The administration of labels within second-level domains is delegated to an organization, which creates a master name file zone and operates a nameserver for resolution of names within the delegated zone domain. For example, Example Corporation registers the label example in the .com TLD. Within the domain example.com, the company can assign names and bind addresses to resources in its network using resource records. NS records identify name servers (resolvers). A and AAAA (quad-A) records associate IPv4 and IPv6 addresses to ASCII-encoded names, such as server33.example.com and pc23.lab.example.com. In addition to DNS names that resolve to individual hosts (using DNS A or AAAA records), DNS SRV resource records can be used to associate names with services.

The Internet uniform resource indicator (URI) or uniform resource locator (URL) [3] is an identifier that identifies an abstract or physical resource. The generic form of a URI/URL is:

```
scheme:user@domain
```

URIs allow different types of resource identifiers to be used in the same context, regardless of the way the resources are accessed. URIs are hierarchical and begin with a scheme name that refers to a specification for assigning identifiers. The next component identifies the authority of a name space (domain name), followed by path or subdelegation components that collectively uniquely identify a resource, as illustrated in these examples from RFC 3986 :

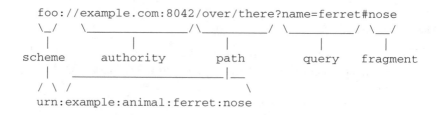

HTTP URIs and SIP URIs both fit this scheme.

11.3 Namespace Management in SIP

Identity assertion and verification of SIP users (or UA endpoint devices) within a single domain can be provided using the same types of user account management applications and databases as organizations today use for data applications. Domain administrators assure that each user (or UA endpoint device) within a domain is assigned a unique user name and credentials appropriate for the authentication method employed in the domain. Organizations will in general manage names and identities for other SIP-aware elements including SIP proxies and SIP application servers. Depending on the authentication method used, SIP UAs may require preconfiguration of names, credentials and addresses of SIP proxies. Some implementations may utilize static configuration files, while others may incorporate SIP configuration information into DHCP or an active directory policy that is uploaded to the SIP UA upon successful network logon. Similarly, SIP proxies and SIP application servers may require preconfiguration of policy (as an authorized URI list) and authentication information, such as a server certificate, approved TLS ciphersuites as well as names, credentials) and addresses of other SIP proxies and the URI schema and authorities (SIP domains that can be reached via each SIP proxy).

11.3.1 URI Authentication

Upon completion of a successful SIP registration by a UA, the SIP server stores the Contact URI extracted from the REGISTER message. Any incoming request to that user's URI can now be routed to that Contact URI address. To protect information exchanged in the registration process, SIP registration can be performed over a TLS connection (generally, any security protocol with equivalent security services) that is kept open and reused for any incoming requests. That is, the TLS connection opened for registration is used for subsequent incoming and outgoing SIP requests for the UA. This is shown in Figure 11.1.

Having authenticated the user, the SIP server can filter incoming requests over this connection. If the `From` header in any incoming request does not match the authenticated URI, the proxy server can reject the request with a 403 `From Identity Not Valid` response as shown in Figure 11.2. The UAs in the domain can be configured to only trust the proxy server to authenticate requests. Any attempt to bypass the proxy servers is rejected with a `305 Use Proxy Server` response. Although a SIP request with forged SIP headers (such as `Via` header fields) can be created, the UA can determine that the certificate is not that of the proxy server and reject the request.

When a SIP domain routes all SIP registrations and requests through a SIP Server, all UAs within the SIP domain trust that the `From` URI represents the authenticated identity, that is, an authorized user in this domain. The SIP domain is a locus of trust that can be extended via the SIP server to other domains. This is shown in Figure 11.3. The assumption the SIP proxy in the second domain makes in this topology is "if I can verify that a SIP request has

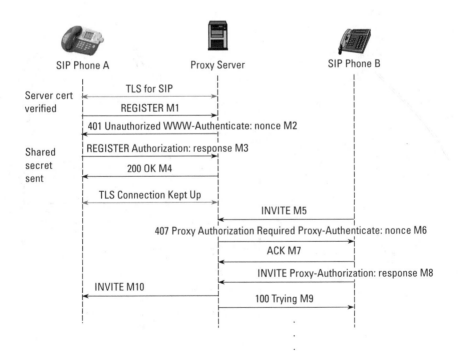

Figure 11.1 Registration and incoming request identity in a single domain.

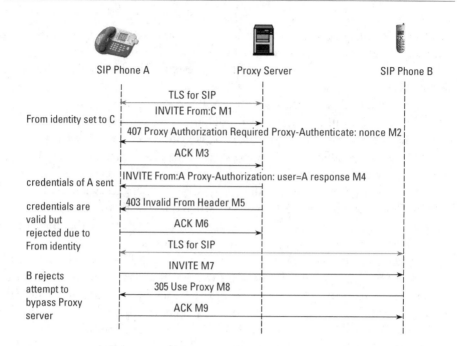

Figure 11.2 From header validation within a domain.

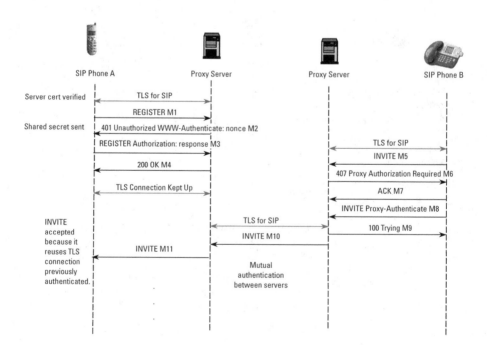

Figure 11.3 Single-hop interdomain authentication.

come directly and securely from the SIP proxy in the first domain, and I trust that the SIP proxy in the first domain is verifying `From` URIs within its local SIP domain, then I can trust URIs from this domain." The SIP proxy in the second domain satisfies the first of these conditions by utilizing mutual certificate-based authentication in TLS, and the second of these conditions by utilizing a TLS cipherset that provides integrity protection to assure that the SIP request and in particular the contents of the `From` header field are delivered without alteration.

This approach has a major drawback. Only the first hop SIP proxy from the domain is able to verify the identity directly: all other parties in the SIP request path must trust this SIP proxy. However, only the SIP proxy in the second domain can directly verify the certificate of the first domain server.

For many VoIP applications, extending a locus of trust in this manner is insufficient. Two other approaches exist. One is based on transitive trust, and the other uses cryptographic signature mechanisms.

11.4   Trust Domains for Asserted Identity

The SIP `P-Asserted-Identity` [4] mechanism emulates the trust and identity model of the PSTN in a SIP VoIP network. It allows a SIP server that has authenticated a user to insert this user's identity information in a `P-Asserted-Identity` SIP header field. A SIP server receiving a request containing an asserted identity bases its decision to accept or refute a `P-Asserted-Identity` on the trust relationship it has with the SIP server from which it received the request. Simply put, a SIP server will trust `P-Asserted-Identities` from a set of trusted SIP servers and no others. The administrators of SIP domains determine the set of trusted SIP servers and thus the trust boundaries of the extended trust domain. Trust in the extended SIP domain is transitive: If a SIP server trusts last hop that processed a SIP request, the asserted identity is kept and can be passed on within the domain of trust. If the last hop that processed the SIP request is not trusted, the asserted identity is ignored and is removed from the request before forwarding within the domain. The `From` URI in the SIP request is thus the only information that is passed, and local policy at the destination SIP network dictates whether the call is accepted or declined. To illustrate this application of transitive trust, consider the trust relationships and the policies they represent with respect to `P-Asserted-Identity` shown in Figure 11.4. In Figure 11.4, A trusts B and C, but D only trusts C. E is not trusted by A, B, C, or D. As such, a request with an asserted identity routed from A to B to C will retain the identity asserted by A. A request routed from A to D will not retain the asserted identity. However, the same request routed from A to C to D will retain the asserted identity,

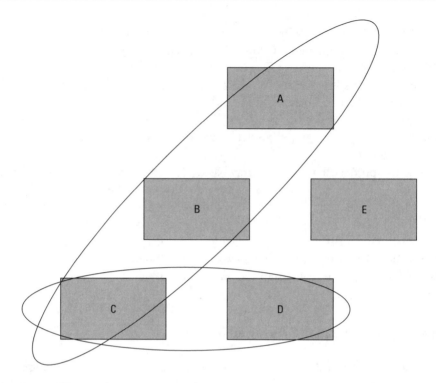

Figure 11.4 SIP trust domain relationships.

since the request has not crossed any trust boundaries. This dependency on the request path, rather than the identity of the party making the assertion shows the underlying weakness of this mechanism. Better identity assertion mechanisms, discussed in subsequent sections, overcome this limitation.

Spec(T) describes a way to assert identity of authenticated users across a network of trusted SIP servers within a SIP administrative domain. All the SIP servers within the administrative domain must implement the private extensions and comply with the behavior prescribed in the Spec(T) specifications, including:

- User authentication method;

- Security mechanisms for intra-domain communication;

- How membership in the trust domain is determined;

- Default privacy mechanism within the domain.

All the users and UA endpoint devices in the SIP administrative domain explicitly trust SIP servers to correctly and publicly assert the identity of SIP parties, and also trust the servers to withhold the identity from disclosure outside the administrative domain when privacy is requested.

A sample set of policies for a secure Spec(T) is described in the following sections.

A UA in a trust domain is configured to only accept incoming requests through a trusted proxy server and must send all outgoing requests to the same trusted proxy: any request not received in this manner can receive a `305 Use Proxy` redirection response. (This is a local configuration setting in the UA endpoint device.) This device might trust the contents of a `P-Asserted-Identity` header field over the contents of a `From` header when presenting CLID to the user or in making call routing decisions.

The `From` URI is set by the initiator of a SIP request and may not match the authenticated identity of the user. When a UA that is a part of a trust domain makes a request, a SIP proxy server may insert a `P-Asserted-Identity` header field of a proxied SIP request. Should the user wish to keep his or her identity private, the UA can invoke the privacy mechanism described in Section 11.11. This might also be a policy an (extended) enterprise might wish to enforce across its SIP administrative domain. It is also possible that a UA might have multiple identities within the domain. Using the `P-Preferred-Identity` header field, the UA can specify which identity a SIP proxy should assert on its behalf. However, this only applies for multiple identities that have the same authentication credentials.

Within a Spec(T) trust domain, SIP proxy servers must authenticate users within the domain. A proxy server may insert a `P-Asserted-Identity` header field in a request after it has authenticated a user and the request is being proxied within the trust domain. If a SIP request is received from a proxy server outside the trust domain, any `P-Asserted-Identity` header fields will be removed before proxying the request.

UA identity information communicated between proxy servers within a Spec(T) trust domain must be protected from eavesdropping, interception, modification, and replay. Confidentiality and integrity protection measures of security protocols such as TLS or IPSec, with appropriate cipher suites, can satisfy this requirement. Mutual endpoint authentication of proxy servers is also recommended.

An example call flow using asserted identity within a Spec(T) trust domain is shown in Figure 11.5.

Some organizations conclude that risk is manageable when transitive trust models like Spec(T) are employed within a single SIP administrative domain. Extended enterprises and more ubiquitous SIP communications may struggle with the risk generally associated with chains of trust, which are only as reliable as the weakest link in the chain. Other organizations may require more reliable interdomain SIP identity verification methods.

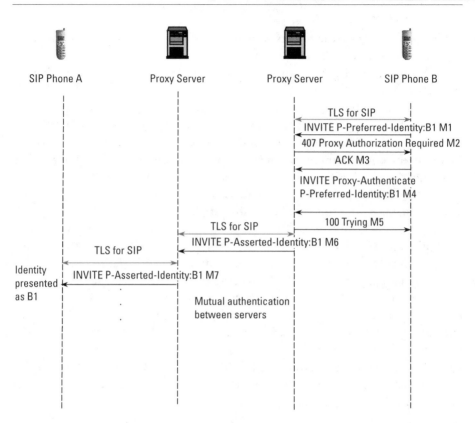

Figure 11.5 P-asserted identity within a Spec(T) trust domain.

11.5   Interdomain SIP Identity

Interdomain SIP identity is a more difficult problem to solve than the single domain case. A single trust domain is not assured and, in fact, will be the exception rather than the rule, so a mechanism is defined through which identity assertions can be verified and traced to the source or originating party. SIP users and proxies can apply policy to this identity information based on who is making the assertion and how this is proven. Two mechanisms for interdomain SIP identity assertions, SIP authenticated identity body and enhanced SIP identity, attempt to solve this problem.

In both mechanisms, the assertion is made by an authoritative SIP server in the domain. Authoritative, in this case, means that the SIP server is responsible for performing authentication and authorization for the domain. In specific terms, for a SIP server to assert an identity within its domain for a given request, the user must present the same credentials in the SIP request as it would in a REGISTER request. Since a REGISTER request binds a user to an identity (address of record), the credentials used here are appropriate for the assertion of the identity.

11.5.1 SIP Authenticated Identity Body (AIB)

The discussion of S/MIME in Chapter 9 shows how a signature computed over the entire tunneled SIP message (`message/sip`) can be used for identity assertion. However, SIP messages can be large, and most of the header fields in the signature have nothing to do with identity and several may change as a request routes through a set of SIP proxies. For example, routing through a set of SIP proxy servers will result in the addition of `Via` header fields and the decrementing of the `Max-Forwards` count header field. As both of these changes are perfectly valid and do not represent a MitM attack, the fact that they change the signature of the message is poses a problem for the receiving UA, which must distinguish these legitimate in-transit header changes from security violations. The *SIP authenticated identity body* (AIB, [5]) computes a signature over the subset of SIP message headers that are relevant to ascertaining the sender's identity and correlating the sender with the request. This subset is known as `message/sipfrag`.

The subset of header fields comprising `message/sipfrag` is chosen to prevent replay and cut-and-paste attacks. For example, the minimum set of `From`, `Date`, `Call-ID`, and `Contact` would provide reasonable assurance against these kinds of attacks. The set may also contain the `To` and `CSeq` header fields for additional protection. If the optional `CSeq` header field is included, then a new AIB must be generated and validated for each request sent during a session, as each new request will have an incremented `CSeq` value. If the `CSeq` value is not used, then the AIB can be reused within the same session. However, it can not be used across sessions due to the `Call-ID` field.

For example, consider the `INVITE` with AIB from RFC 3893:

```
INVITE sip:bob@example.net SIP/2.0
Via: SIP/2.0/UDP pc33.example.com;branch=z9hG4bKnashds8
To: Bob <sip:bob@example.net>
From: Alice <sip:alice@example.com>;tag=1928301774
Call-ID: a84b4c76e66710
CSeq: 314159 INVITE
Max-Forwards: 70
Date: Thu, 21 Feb 2002 13:02:03 GMT
Contact: <sip:alice@pc33.example.com>
Content-Type: multipart/mixed; boundary=unique-boundary-1

—unique-boundary-1
Content-Type: application/sdp
```

```
Content-Length: 147

v=0
o=UserA 2890844526 2890844526 IN IP4 example.com
s=Session SDP
c=IN IP4 pc33.example.com
t=0 0
m=audio 49172 RTP/AVP 0
a=rtpmap:0 PCMU/8000

—unique-boundary-1
Content-Type: multipart/signed;
protocol="application/pkcs7-signature";
micalg=sha1; boundary=boundary42
Content-Length: 608

—boundary42
Content-Type: message/sipfrag
Content-Disposition: aib; handling=optional
From: Alice <sip:alice@example.com>; tag=1928301774
To: Bob <sip:bob@example.net>
Contact: <sip:alice@pc33.example.com>
Date: Thu, 21 Feb 2002 13:02:03 GMT
Call-ID: a84b4c76e66710
CSeq: 314159 INVITE

—boundary42
Content-Type: application/pkcs7-signature; name=smime.p7s
Content-Transfer-Encoding: base64
Content-Disposition: attachment; filename=smime.p7s;
handling=required

ghyHhHUujhJhjH77n8HHGTrfvbnj756tbB9HG4VQpfyF467GhIGfHfYT6
4VQpfyF467GhIGfHfYT6jH77n8HHGghyHhHUujhJh756tbB9HGTrfvbnj
n8HHGTrfvhJhjH776tbB9HG4VQbnj7567GhIGfHfYT6ghyHhHUujpfyF4
7GhIGfHfYT64VQbnj756

—boundary42—
—unique-boundary-1—
```

The AIB signature in this example covers the From, To, Contact, Date, Call-ID, and CSeq header fields.

11.5.2 Enhanced SIP Identity

The enforcement of enhanced SIP identity in interdomain signaling begins at the SIP server, as it does in the single domain case. An authoritative SIP server authenticates all users in the SIP administrative domain by verifying the URI in the `From` header field of the request. Authenticated requests are forwarded through the SIP server to proxies or gateways outside the domain. Before doing so, however, the authoritative SIP server signs the request with its private key and appends this signature to the request in a special, additional header field, the `Identity` header field [6].

SIP does not assume a PKI infrastructure—an additional header field can also be appended to assist in the retrieval of a public certificate. The value of `Identity-Info` provides a URI which can be used to retrieve the certificate of the server. `Identity-Info` is included in the integrity protection computation by the authoritative SIP server prior to signing the request.

When the request is received in another trust domain, the processing SIP server can check the certificate chain and verify the authoritative SIP Server's signature. Every proxy receiving the request can verify the identity, then forward it on to another proxy server or UA, effectively solving the trust problem created when chained proxies rely solely on the first SIP proxy to verify identities "locally." This solution provides end-to-end asserted identity. A UA multiple hops away from the authoritative SIP server can determine who made the identity assertion by checking who signed the `Identity` header field. The UA can accept or reject the request based on the explicit trust relationship the UA's local SIP administrative domain establishes with the SIP administrative domain where the call originated. Although the receiving UA has no way of determining if secure transport has been used at each hop, or if all the previous proxies are trustworthy, the UA can be confident that the asserted identity is authentic if it is able to verify the authoritative SIP server's signature in the `Identity` field.

The example SIP request below contains `Identity` and `Identity-Info` header fields which were inserted by the proxy2.example.com SIP server:

```
INVITE sips:bob@biloxi.exmple.org SIP/2.0
Via: SIP/2.0/TLS pc33.atlanta.example.com;branch=z9hG4bs8
Via: SIP/2.0/TLS proxy2.example.com;branch=z9hG4bKna432
To: Bob <sips:bob@biloxi.example.org>
From: Alice <sips:alice@atlanta.example.com>;tag=932sdf28
Identity:
  rtwretwertikp324515p08rgpojq3459ui423jqewrojojertjrt
  Fkj5o098145oj234tkjerrgoiueroitowi34ritjewrtjlerjt
  ;alg=rsa-sha1
Indentity-Info: <https://www.atlanta.example.com/sipcert>
Call-ID: 4b4kfc76e66710
CSeq: 141594 INVITE
Max-Forwards: 69
Date: Thu, 21 Feb 2005 13:02:03 GMT
Contact: <sips:alice@pc33.atlanta.example.com>
```

```
Content-Type: application/sdp
Content-Length: 147

v=0
o=- 2890844526 2890844526 IN IP4 pc33.atlanta.example.com
s=Session SDP
c=IN IP4 pc33.atlanta.example.com
t=0 0
m=audio 49172 RTP/AVP 0
a=rtpmap:0 PCMU/8000
```

The set of SIP header fields over which the signature is computed is presented in Table 11.1. The use of the `Identity` and `Identity-Info` header field is shown in Figure 11.6. The `Identity` header field is applicable to any SIP request method with the exception of `CANCEL`, which may not use this mechanism due to its hop-by-hop nature.

Since the message body is covered, this means that SDP which may include a MIKEY key management message, or the contents of a `NOTIFY` message body (for example, a certificate fetched using the SIP certificate store) has integrity protection from this identity mechanism.

Computing the signature over the `To` header field provides integrity protection against a cut and paste attack in which a valid `Identity` signature is used with a request sent to another UA. The `Call-ID` uniquely identifies the session while the `CSeq` count identifies the transaction. Within a dialog, each request generated by a UA will have an increasing `CSeq` number. The `Date` is provided as a test of staleness, that is, an old `Identity` header field is not being replayed by an attacker. The `Contact` URI identifies the actual device making the request.

Table 11.1
Set of SIP Header Fields Covered by the Identity Header Field

Header	Reason
From URI	Actual identity being asserted
To URI	The destination of the request
Call-ID	Identifies the actual dialog
CSeq	Identifies the actual transaction, each transaction will have an incremented CSeq count
Date	Timestamp
Contact URI	Actual UA making the request
message body	The message body payload

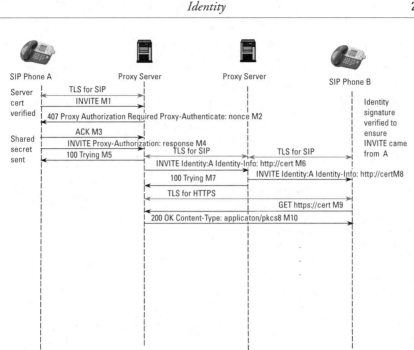

Figure 11.6 Use of Identity and Identity-Info header field.

Some SIP requests (INVITE, SUBSCRIBE, NOTIFY, UPDATE, and REFER) must have a Contact header field. Other requests (OPTIONS, REGISTER, ACK, and INFO) may or may not contain a Contact header field. Certain requests (BYE, MESSAGE, PRACK, and PUBLISH) will never have this header field. If the Contact header field is not present, no value will be used in the hash.

The hash algorithm is specified in the algorithm (alg) parameter in the Identity-Info header field. The only currently defined algorithm is rsa-sha1 which refers to SHA-1 with RSA encryption using a minimum key length of 1024 bits. The Identity-Info contains a HTTP or HTTPS URI which resolves to an application/pkix-cert resource.

A UA or proxy that requires the presence of an Identity header field can fail a request which does not contain one by sending out a 428 Use Identity Header Field response, as shown in Figure 11.7.

A UA or proxy that is unable to access the URI in the Identity-Info header field to validate the signature can return a 436 Bad Identity-Info Header Field response. A UA or proxy that is able to access the certificate in the Identity-Info but is unable to validate the certificate can return a 437 Unsupported Certificate response.

The use of call forwarding makes response identity a more difficult problem. The Identity header field in a request validates the From URI. If a

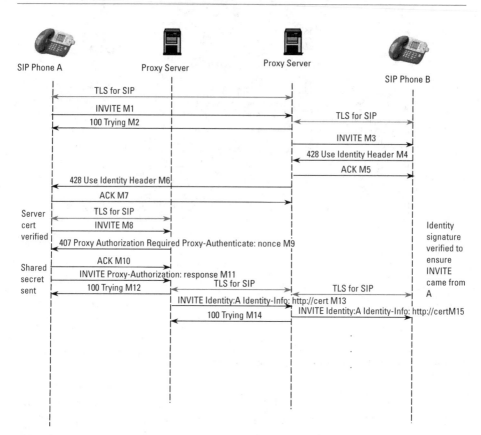

Figure 11.7 UA requesting the use of the identity header field.

request is forwarded, the identity of the party that answers the request may be different from the To URI. As such, the principles of request identity cannot be applied to responses in SIP. Until a robust response identity mechanism is developed, conventional methods of authenticating SIP responses must be used, as discussed in Chapter 9.

One solution to this problem is to use redirection (3xx) instead of forwarding. This solution is shown in Figure 11.8. Besides making response identity easier, it also solves problems if a body in the request was encrypted with the public key of the intended recipient (To URI), but the request is answered by another party. A redirection allows the requestor to fetch the public key of the redirected party then resend the request. Note that the History-Info header field [7] can be used to validate that this redirection is authorized.

In addition to the problems associated with call forwarding already mentioned, intermediary systems such as proxy and redirection servers may generate SIP responses on behalf of a UA. Some work is underway to allow the generator

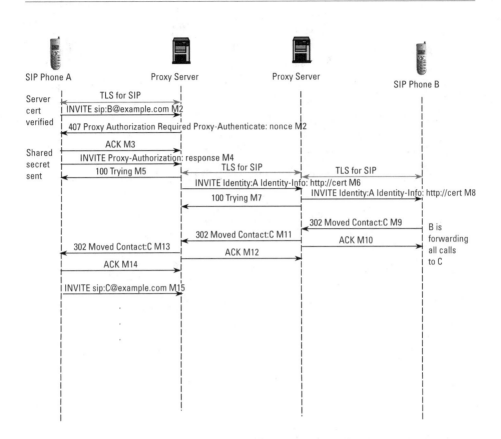

Figure 11.8 Forwarding implemented Using redirection.

of a response to provide its identity, sign the response, and provide a pointer to its certificate.

This enhanced identity scheme can also coexist with the transitive trust mechanism described in Section 11.4. For example, when a request containing an identity assertion crosses a trust boundary, the receiving SIP proxy can validate the identity and signature, then add a P-Asserted-Identity header field for use within its own trust domain. This is shown in Figure 11.9.

11.6   SIP Certificates Service

A VoIP system using SIP can utilize a public key infrastructure (PKI) to obtain and use certificates for authentication, identity, and signatures. The use of certificates by SIP servers such as proxy servers, redirect servers, and registrar servers was discussed in Chapter 9. The use of certificates by SIP servers is identical to

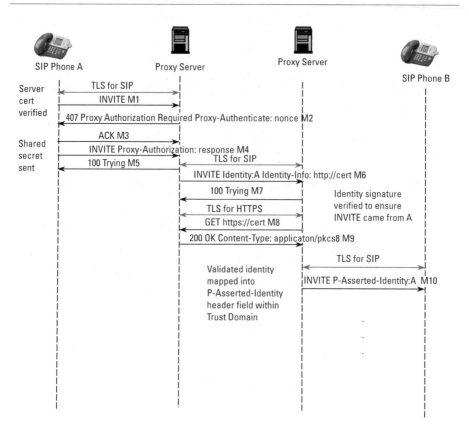

Figure 11.9 Mapping enhanced identity in a trust domain.

their usage in normal secure web (HTTPS) and electronic commerce applications. This is by design, as this model has proven to be scalable for web servers and e-commerce servers, and will likely be scalable for SIP servers as well. The use of client certificates for web servers is problematic.

Some enterprises have deployed client certificates in user's laptops. These experiences expose the following problems:

1. A seamless, global PKI infrastructure is unavailable. While organizations can obtain certificates for all their employees from a public CA such as Thawte or Verisign, the CAs used for client certificate issuance are typically operated as an enterprise service, which makes certificate management simpler but makes the certificates usable only within the enterprise. If the certificates were issued by a publicly recognized and trusted CA (one whose root certificate is built into most web browsers), the inverse is true: clients certificates can be used for any Internet application, but the enterprise must involve the public CA each time a

user joins or leaves the *extended* enterprise, and must work within the credentialing framework used by the CA's registration authority.

2. Certificate management is resource-consuming. Installation of certificates on user devices, training of users, protection of private keys, and other certificate maintenance issues can increase helpdesk costs. Many PKI-enabled organizations store certificates on separate devices, such as a smart cards or USB dongles. While simplifying some aspects of certificate administration, this introduces problems of managing these devices themselves.

3. The use of a single certificate across a number of applications and usages has proven to be problematic. This is an example of the "single signon" problem which has proven to be difficult to solve. Many enterprises that have attempted this have gone back to separate authentication systems for separate systems.

4. Management of revoked or expired certificates has proven very troublesome. Certificate revocation lists (CRLs) are particularly inefficient in real-time scenarios.

The use of client certificates by SIP UAs is perhaps even more problematic. However, there seems to be considerable value in, and hence significant motivation for implementing a SIP certificate service.

Consider now the deployment issues associated with putting certificates into SIP UAs for the purposes of secure interdomain communications.

1. Having an enterprise or service provider serve as the CA will be of little value in the commonly compelling case of interdomain communication in which there has been no previous communication between endpoints.

2. A user may have multiple SIP devices such as a SIP phone, PDA, cell phone, and WiFi phone. Since the same user (same AOR) will be using all these devices, the same certificate must be installed (or plugged into) all these devices, a difficult administration task over many different devices. Having different certificates in each of these devices is not feasible unless the initiator of communication knew beforehand which device communication was to be established with.

3. Installing and managing certificates on endpoint devices with limited storage is harder than in a PC or laptop.

4. The SIP user population will behave more like cellular subscribers than enterprise laptop users. This will be especially true for public SIP operators, who must anticipate a high incidence of subscriber "adds, drops,

and changes." SIP operators will need to process service orders in real time, which may involve, for example, certificate suspension or revocation, based on service portability, or failure to pay bill, for instance.

5. The sheer volume of client certificates SIP operators must manage is daunting. Including wireless networks, tens of millions of SIP UAs might need client certificates in the coming years.

6. Creating trust relationships across multiple certificate authorities is a currently unsolved problem. The continued failure to get users of email to obtain and use (S/MIME) certificates is a particularly troubling insight into the problems faced by SIP.

As a result, the IETF has sought an alternative to the public and private CA models. The SACRED (Securely Available Credentials) Working Group in the IETF has developed a framework [8] that potentially solves problems 1 through 3 as shown in the list above.

SACRED was developed to support user credential access and management for roaming and mobile environments. SACRED provides a protocol framework that allows a user to securely authenticate to a *credential server* using a strong password protocol. Once authenticated, a user can upload or download user credentials (digital certificates) that can subsequently be used for authentication in SIP communications. SACRED thus empowers the user rather than a private enterprise with the ability and responsibility of making his certificate widely available (as indicated in problem 1 listed above). The SACRED framework assumes a network architecture where the credential server is conceptually distinct from the secure repository for credentials, known as the credential store. In principle, a user can run the (conceptual) SACRED client-server protocol from any of possibly many IP-enabled endpoint devices he operates, thus solving the "one client certificate on many endpoint devices" and the "managing certificates on many endpoint devices" problems (2 and 3 above). The SACRED framework only defines the client-credential server exchange. Where credential server and store are distinct systems, SACRED assumes that these systems can use existing protocols such as TLS, LDAP, or secure LDAP to upload and retrieve user credentials.

Note that this approach is inherently less secure than conventional means of installing and moving certificates, and this is *by design*. SACRED protocols by themselves cannot solve the problem of managing trust relationships that continues to hamstring PKI (problem 6 above) and so are not suitable for interdomain credential access and management. However, they are suitable for the establishment of real-time sessions between users.

The SACRED framework defines a credential server as a secure repository where client credentials (digital certificates, shared secret keys, unsigned public/private keypairs) are stored. SACRED defines three operations: credential upload (PUT), credential download (GET), and credential removal (DELETE). The steps in a credential upload are conceptually described as follows:

1. The client and server mutually authenticate each other using a strong password protocol and negotiate a session level encryption key.

2. The client creates a data object consisting of his user credential and a credential format identifier, a field that identifies the type and encoding of a credential. A PKCS #15 encoded X.509 certificate meets these criteria. This secure credential file should be privacy and integrity protected. The client uploads the secure credential file to the credential server over an encrypted tunnel to the certificate server. The session encryption key exchanged during mutual authentication serves as the secret key for the symmetric encryption selected for securing this tunnel.

3. The server responds with a success or error condition.

Credential download follows similar steps. Following mutual authentication and session level key exchange, the client issues GET requests and the credential server responds by sending secure credential files to the client over an encrypted tunnel. A client issues a DELETE to remove a secure credential file in the same manner.

The SIP certificate service is an implementation of a SACRED Client-to-Credential Server protocol. The basic operation is shown in Figure 11.10. Note that the credential server is colocated with the proxy server. This is but one example of how these functions might appear in actual deployments. In this flow, the UA uses the PUBLISH method (messages M1 through M4) to perform the credential upload operation. The certificates are self-generated and self-signed by the user. The UA uses SUBSCRIBE and NOTIFY (messages M5 through M10) for credential download. The UA satisfies step 1 of the SACRED Client-to-Certificate-Server protocol by opening of a TLS session to a certificate service. The UA authenticates the certificate server by verifying its certificate. The server authenticates the UA by issuing a digest challenge. The client returns a MD5 hash of a secret it shares with the server thus satisfying SACRED's mutual authentication requirement. The MD5 hash is known as the account password in the SACRED framework. The client and certificate server negotiate a session key during the TLS handshake, as described in Chapter 6.

The client and certificate server satisfy step 2 of SACRED requirements by using the TLS Application Data Protocol to provide encrypted and signed transport for credentials and certificates carried in the PUBLISH request. The PUBLISH request contains a multipart/MIME message body with an

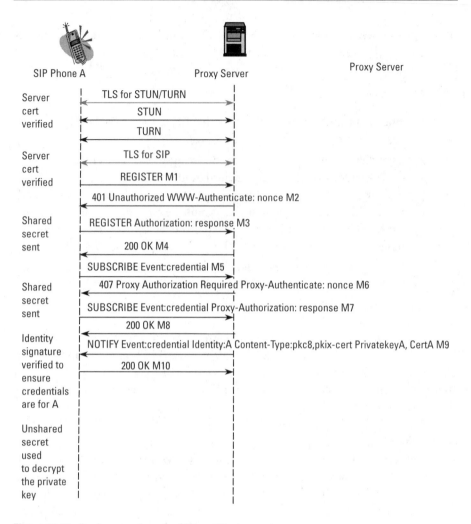

Figure 11.10 Basic operation of a SIP certificate service.

application/pkc8 MIME type, defined in [9] as a DER encoded PKCS#8 [10] private key credential and an application/pkix-cert MIME type, defined in [11] as an X.509 certificate. The PUBLISH contains a Event: credential header field that identifies the event package for credential.

The response to the PUBLISH request, is a 200 OK for a success, satisfies the step 3 requirements of SACRED. Note that the request is sent to the AOR URI of the user. An example of a PUBLISH request is shown below:

```
PUBLISH sip:bob@206.65.230.170:64030 SIP/2.0
To: "Bob"<sip:bob@voiptheworld.net>;tag=7a6ac719
From: <sip:alice@ipislands.com>;tag=f653770c
```

```
Via: SIP/2.0/UDP 206.65.230.134;branch=z9hG4bK3281
Via: SIP/2.0/UDP 206.65.230.135;branch=z9hG4bK3281
Via: SIP/2.0/UDP
206.65.230.170:64032;branch=z9bK-d86b-1—d843-;rport=64032
Call-ID: c360da332f48a04c@U01QX0RFTU8x
CSeq: 2 NOTIFY
Contact: <sip:alice@206.65.230.170:64032>
Max-Forwards: 68
Content-Type: multipart/mixed;boundary=boundary
Subscription-State: active;expires=3000
Event: credential
Content-Disposition: signal
Content-Length: 543

--boundary
Content-ID: 123
Content-Type: application/pkix-cert
```

[CertA]

```
--boundary
Content-ID: 456
Content-Type: application/pkcs8
```

[PrivateKeyA]

```
--boundary
```

The credential itself is stored on the certificate service in encrypted form, and requires a credential password, as described in the SACRED framewᵣrk. This credential password is never sent over the Internet, to prevent disclosure or theft of the credential. If the certificate store or the communication between the certificate store and the client is compromised, the credential will still not be usable to the attacker unless the password is also known.

Client and certificate server perform certificate retrieval over a TLS connection in the same manner. The UA issues a SUBSCRIBE request, containing an Event: credential header field, to authorize the certificate service to issue notifications whenever his credential changes. The certificate server issues an initial NOTIFY over the TLS connection to provide the current state (the current credential). A user who simultaneously operates multiple UAs can retrieve multiple copies of the credential and use the private key on each of the devices, as shown in Figure 11.11.

This same approach can be implemented as a general certificate retrieval mechanism in the absence of a PKI infrastructure. UAs can generate self signed certificates and upload these to the certificate service. Using the same

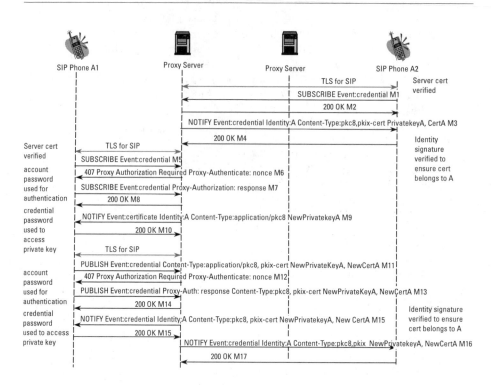

Figure 11.11 Multiple UAs sharing the same credential.

mechanisms as for credential storage and retrieval, certificate service can store and hand out the certificates and public keys of the UAs. This is shown in Figure 11.12. In M1 - M4, another UA outside the domain of SIP phone A wants to know A's public key. The UA sends a SUBSCRIBE request to A's AOR URI which is routed to the certificate service in SIP phone A's domain. The request carries an Event:certificate and Accept-Content: application/pkix-cert which is a MIME type for carrying a X.509 certificate as described in RFC 2585.

In messages M5 - M8, a UA generates and uploads a new private key/public key pair to the certificate service. SIP phone B still has an active subscription to SIP phone A's public key, so SIP phone B receives an immediate NOTIFY which contains SIP phone A's new certificate. This NOTIFY contains an application/pkix-cert MIME body.

For UAs that have long-lived associations (for example, they might be on each other's buddy list), this subscription would be continuous. The UAs would each have the most recent copy of each other's certificate.

For UAs that do not have long-lived associations or have never communicated before, the certificate can be fetched when it is needed. This will introduce a slight additional delay in establishing a session as in some cases the certificate

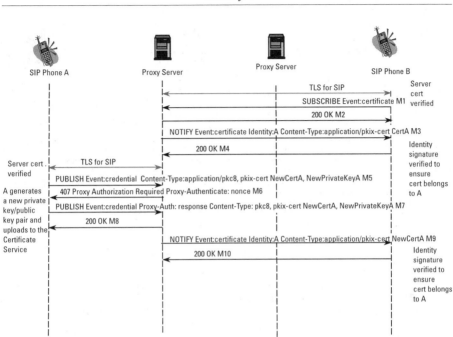

Figure 11.12 Certificate service storing and retrieving public keys.

may have to be fetched before initiating the session. As a result, the certificate is never cached.

The only caching issue is how to satisfy persistent certificate subscriptions. If the refresh interval for the subscription is, for example, four hours, then the UA effectively is caching the certificate for this duration. If the certificate is expired within those four hours, a new NOTIFY will be sent. However, an attacker might be able to prevent that NOTIFY from being sent or delivered. An example of real time use of a Certificate service is shown in Figure 11.13.

11.7   Other Asserted Identity Methods

Other methods for asserting identity are being studied. We briefly consider these in the following sections. Many of these methods are proposals and have no formal status in standards communities at this time.

11.7.1 Secure Assertion Markup Language

The Secure Assertion Markup Language, or SAML [12], is an XML encoded protocol for transferring identity and authorization information. SAML has been

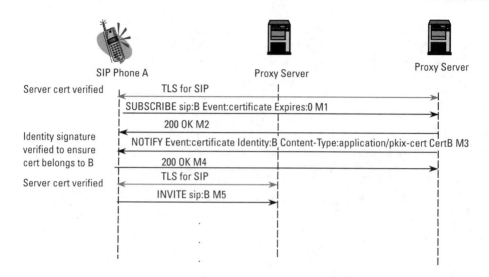

Figure 11.13 Real-time use of a certificate service.

developed by OASIS (Organization for the Advancement of Structured Information Standards) [13], a committee that develops standards for e-commerce.

SAML has two parts: SAML assertions, which describes the document format, and the SAML protocol, which defines a mechanism for querying and retrieving SAML information. SAML transport can be HTTP, SOAP, or other protocols.

A proposed use of SAML by SIP to assert identity is described in [14]. An example SAML message from this document is shown below:

```
<saml:Assertion
xmlns:saml="urn:oasis:names:tc:SAML:1.0:assertion"
MajorVersion="1"
MinorVersion="1"
AssertionID="P1YaAz/tP6U/fsw/xA+jax5TPxQ="
Issuer="www.example.com"
IssueInstant="2004-06-28T17:15:32.753Z">
        <saml:Conditions
        NotBefore="2004-06-28T17:10:32.753Z"
        NotOnOrAfter="2004-06-28T17:20:32.753Z" />
        <saml:AuthenticationStatement
AuthenticationMethod="urn:ietf:rfc:3075"
AuthenticationInstant="2004-06-28T17:15:11.706Z">
        <saml:Subject>
        <saml:NameIdentifier>
        NameQualifier=alice@example.com
        Format="urn:oasis:names:tc:SAML:1.1:nameid-
```

```
          format:emailAddress">uid=alice
          </saml:NameIdentifier>
          <saml:SubjectConfirmation>
                  <saml:ConfirmationMethod>
                  urn:oasis:names:tc:SAML:1.0:
                  cm:SIP-artifact-01
                  </saml:ConfirmationMethod>
          </saml:SubjectConfirmation>
          </saml:Subject>
     </saml:AuthenticationStatement>
  </saml:Assertion>
```

This example shows an `application/saml+xml` [15] message body that could be used as part of a trait based authorization system for SIP [16].

11.7.2 Open Settlements Protocol and VoIP

Open Settlements Protocol (OSP) [17] is a protocol used in some VoIP systems to provide provisioning, authentication, and accounting services. An architecture describing the use of OSP with SIP is shown in Figure 11.14.

An extension to SIP to carry an OSP Authorization token [18] has been proposed to the IETF. As shown in Figure 11.14, an element such as a proxy requests a token from the OSP server, and includes the token in the `INVITE` routed over the Internet. The destination proxy server can validate the token with the OSP server for authentication and billing purposes.

11.7.3 H.323 Identity

The H.323 protocol has a number of identifiers. The simplest is the H.323 alias, which is effectively just a username within a domain. H.323 systems can also utilize E.164 numbers and private numbering schemes.

H.323 also has a URI scheme defined, known as Annex O and published as RFC 3508 [19]. A H.323 URL begins with `h323:` and can contain just a H.323 alias or a complete name. The user part contains the alias address while the host part resolves to the H.323 Gatekeeper or border element that can route the request.

In general, identity is checked and asserted between H.323 gatekeepers using mechanisms similar to those in the trust domains described in Section 11.4.

11.7.4 Third Party Identity and Referred-By

The `REFER` method [20] is a SIP method to request another UA act upon a URI. When used with a SIP or SIPS URI, the resulting action will be to send an

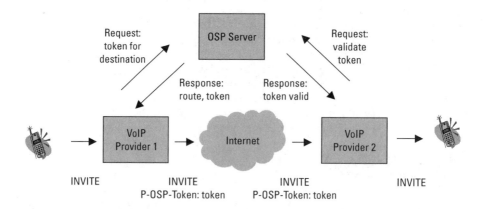

Figure 11.14 Open Settlements Protocol and SIP with a clearinghouse.

INVITE to the destination to establish a new session. The resulting session is being established at the request of a third party. The Referred-By mechanism [21] is a way in which the third party can provide a cryptographically verified token to indicate this.

The information in this header field is encrypted to protect against tampering by the third party undertaking the REFER.

A referrer wishing to provide this information includes a Referred-By header field in the REFER. The header field references a message body which is an Authenticated Identity Body.

In Figure 11.15, an unauthenticated Referred-By is included in a REFER triggered request. The recipient requests that the Referred-By be authenticated using a 429 Provide Referrer Identity response. The REFER is then resent with the Referred-By and AIB body.

11.8 Privacy

Thus far, we have has discussed identity and ways in which identity can be asserted and secured through a SIP VoIP network. We now consider mechanisms to keep identity private.

A SIP message contains a fair amount of information that a UA may wish to treat as private or sensitive. This includes both SIP URIs present in From and Contact header fields and IP Addresses present in Via, Contact, Call-ID, and SDP.

A UA can influence whether information conveyed in SIP header fields is treated as sensitive. For example, a UA seeking privacy can refrain from including its AOR URI in a From header field and instead use an anonymous header field such as:

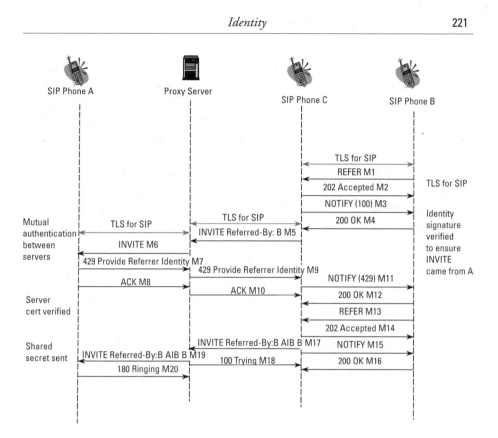

Figure 11.15 Requesting Authenticated Referred-By.

```
From "Anonymous" <sip:anonymous@anonymous.invalid>
```

Similarly, a UA can encode a pseudorandom string in a `Call-ID` field instead of an IP address. However, a UA cannot always provide complete privacy and still be able to establish communication with another user. One approach is to employ some kind of anonymizing or privacy-aware intermediary.

One approach to offering privacy and anonymity is to use a back-to-back user agent (B2BUA) which can act as a proxy for both the signaling and media and in particular remove various types of identifying information. This form of proxy is similar to an SMTP proxy commonly found on application proxy firewalls. SMTP proxies are commonly used to "normalize" e-mail addresses to hide the details of mail service within an organization. Consider a multinational organization example.com that uses country codes' fourth-level domain name labels as a naming convention for SMTP servers (`us.mail.example.com`, `uk.mail.example.com`). An SMTP proxy can modify mail headers before email is forwarded outside the organization so that all mail appears to come from the generic `mail.example.com` to prevent mail traffic analysis. SMTP proxies also support methods for hiding IP addresses and rewriting other mail headers from messages before they are forwarded to external SMTP mail hosts.

The `Privacy` header field [22] provides similar methods and enables a UA to request various levels of privacy from servers within the trust boundary. The levels that can be communicated in the header field are summarized in Table 11.2.

A B2BUA is only one of several ways to implement UA privacy services. For example, transport addresses obtained using TURN [23] can be used to anonymously receive media.

The message below shows how a UA can utilize TURN for privacy:

```
INVITE sips:bob@biloxi.exmple.org SIP/2.0
Via: SIP/2.0/TLS 166.43.32.1:33212;branch=z9hG4bKnashds8
To: Bob <sip:bob@biloxi.example.org>
From: Anonymous <sip:anonymous@anonymous.invalid>;tag=192
Call-ID: a84b4c76e66710
CSeq: 314159 INVITE
Max-Forwards: 70
Date: Thu, 21 Feb 2002 13:02:03 GMT
Contact: <sips:34345234454@166.43.32.1:34212>
Content-Type: application/sdp
Content-Length: 147

v=0
o=UserA 2890844526 2890844526 IN IP4 166.43.32.1
s=
c=IN IP4 166.43.32.1
t=0 0
m=audio 34232 RTP/AVP 0
a=rtpmap:0 PCMU/8000
```

Table 11.2

SIP Privacy Handling Modes

Mode	Description
Header	Headers such as Via should be hidden/rewritten
Session	SDP and other session related
User	User level privacy between intermediaries
ID	User level privacy requested by a UA
None	No privacy is requested.
Critical	If privacy can not be provided, route request anyway

Using TURN, the TLS transport address 166.43.32.1:33212 is obtained for SIP signaling (and used in the `Via` and `Contact` URIs) while 166.43.32.1:3432 and 166.43.32.1:3433 are obtained for UDP RTP and RTCP traffic and signaled in the SDP.

Note that the "lock down" properties of TURN mean that new TURN allocations would be needed for each SIP session establishment. When TURN is used in this manner, care must be exercised to assure that the IP addresses associated with the TURN server cannot be reverse mapped to an identifying domain.

References

[1] Crocker, D., "Standard for the Format of ARPA Internet Text Messages," RFC 822, August 1982.

[2] Mockapetris, P., "Domain Names—Implementation and Specification," IETF RFC 1035, November 1987.

[3] Berners-Lee, T., R. Fielding, and L. Masinter, "Uniform Resource Identifier (URI): Generic Syntax," RFC 3986, January 2005.

[4] Jennings, C., J. Peterson, and M. Watson, "Private Extensions to the Session Initiation Protocol (SIP) for Asserted Identity within Trusted Networks," RFC 3325, November 2002.

[5] Peterson, J., "Session Initiation Protocol (SIP) Authenticated Identity Body (AIB) Format," RFC 3893, September 2004.

[6] Peterson, J., "Enhancements for Authenticated Identity Management in the Session Initiation Protocol (SIP)", IETF Internet-Draft, draft-ietf-sip-identity-03, work in progress, September 2004.

[7] Barnes, M., "An Extension to the Session Initiation Protocol for Request History Information," IETF Internet-Draft, work in progress, January 2005.

[8] Gustafson, D., M. Just, and M. Nystrom, "Securely Available Credentials (SACRED) - Credential Server Framework," RFC 3760, April 2004.

[9] Jennings, C., and J. Peterson, "Certificate Management Service for SIP," Internet-Draft,draft-ietf-sipping-certs-00, (work in progress), October 2004.

[10] RSA Laboratories, "Private-Key Information Syntax Standard,Version 1.2," PKCS 8, November 1993.

[11] Housley, R., and P. Hoffman, "Internet X.509 Public Key Infrastructure Operational Protocols: FTP and HTTP," RFC 2585, May 1999.

[12] Maler, E., R. Philpott, and P. Mishra, "Assertions and Protocol for the OASIS Security Assertion Markup Language (SAML) V1.1," September 2003.

[13] http://www.oasis-open.org/.

[14] Tschofenig, H., J. Peterson, J. Polk, D. Sicker, and M. Tegnander, "Using SAML for SIP," Internet-Draft, work in progress, October 2004.

[15] Hodges, J., "application/saml+xml Media Type Registration," IETF Internet-Draft, draft-hodges-saml-mediatype-01, work in progress, June 2004.

[16] Peterson, J., "Trait-Based Authorization Requirements for the Session Initiation Protocol (SIP)," IETF Internet-Draft, work in progress, February 2004.

[17] European Telecommunications Standards Institute, "Telecommunications and Internet Protocol Harmonization Over Networks (TIPHON); Open Settlement Protocol (OSP) for Inter-Domain Pricing, Authorization, and Usage Exchange," Technical Specification 101 321. Version 2.1.0.

[18] Johnston, A., D. Rawlins, S. Thomas, and R. Brennan, "Session Initiation Protocol Private Extension for an OSP Authorization Token," IETF Internet-Draft, work in progress, June 2004.

[19] Levin, O., "H.323 Uniform Resource Locator (URL) Scheme Registration," RFC 3508, April 2003.

[20] Sparks, R., "The SIP Refer Method," RFC 3515, April 2003.

[21] Sparks, R., "The Session Initiation Protocol (SIP) Referred-By Mechanism," RFC 3892, September 2004.

[22] Peterson, J., "A Privacy Mechanism for the Session Initiation Protocol (SIP)," RFC 3323, November 2002.

[23] Rosenberg, J., R. Mahy, and C. Huitema, "Traversal Using Relay NAT (TURN)," IETF Internet-Draft, work in progress, February 2005.

12

PSTN Gateway Security

12.1 Introduction

Since the number of VoIP enabled endpoints is likely to be significantly less than the number of public switched telephone networks (PSTN) endpoints for the foreseeable future, interconnection with the PSTN is a key part of any VoIP system. This chapter will focus on some of the issues specific to PSTN interconnection, including gateway security, PSTN (toll) fraud, and telephone number mapping using the DNS.

12.2 PSTN Security Model

Security in the PSTN is primarily achieved through physical isolation rather then cryptographic measures.

Consider PSTN signaling. In the early days of the telephone network, the time division multiplexing (TDM) physical bearer channels also carried tones used for signaling and authentication. Attackers known as "phone phreakers" exploited this multiplexing of signaling and conversations onto a single channel. By mimicking signaling tones, phreakers were able to place free telephone calls. Phreakers quickly developed various tone generation boxes known as "red boxes" and "blue boxes" to manipulate the telephone signaling network.

At the time, long distance calling was particularly expensive, and telephone network operators quickly recognized that left unchecked, phreaking posed a serious threat to revenue. As a countermeasure to phreaking, telecommunications carriers chose to isolate the signaling channel from the bearer channel. Signaling was carried out of band over a physically separate network that was not accessible to users. Access to the Signaling System #7 (SS7) packet-switched network is very tightly controlled and monitored by telecommunication carriers. The telecommunications carriers are confident that physical security measures are sufficient, so neither encryption nor authentication is used in SS7 networks.

As a further security measure against signaling tampering, the PSTN uses different signaling protocols for trusted and untrusted elements. For trusted elements, network to network interfaces (NNI) such as SS7 are used for interconnection. PSTN operators use NNIs within their switch network and in bilateral interconnection with other trusted operators. For untrusted elements, user to network interfaces (UNI) such as integrated services digital network (ISDN) are utilized. It is the responsibility of the PSTN operator to validate information received from a UNI before the information is passed along a NNI. This is shown in Figure 12.1.

PSTN switches generate per-call auditing information known as call detail records (CDRs). These are often analyzed in real time to detect fraudulent activities. Unusual calling patterns result in flagged transactions that are investigated carefully. A delay in a few days or weeks in detecting and shutting down fraudulent activity can make the difference between a small and multimillion dollar loss.

The PSTN network also has protection against denial of service attacks, both malicious and unintentional. For example, television shows, consumer polls, and radio station contests can generate large simultaneous call volume to a single number. The PSTN utilizes a method known as automatic call gapping (AGC) to deal with these scenarios. Higher than typical call volumes are detected and squelched as close to the point of origination by the application of

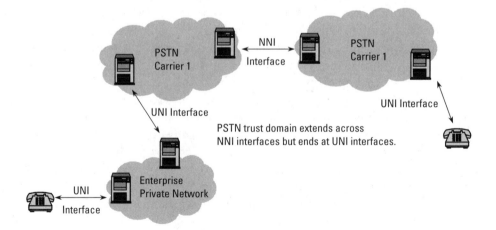

Figure 12.1 UNI and NNI interfaces in the PSTN.

the "busy" or "reorder" treatment. Only a certain number of the calls placed to the gapped number will actually be routed, resulting in traffic squelching.

This approach is also used to limit volume to the PSTN's emergency calling system, known as E911 in North America. E911 uses a very small number of trunks (switch ports) from each telephone switch, which limits the volume before calls reach the selective router and the public service answering point (PSAP), the call center where operators answer emergency calls.

PSTN security can also be effectively mandated and controlled by legislation. The PSTN uses physical facilities and is bound to a particular geographic location, which places elements of the network under various (as state and municipal) jurisdictions. PSTN service providers are licensed and regulated by various levels of government, as, for example, the U.S. Federal Communications Commission, to a much greater degree than the Internet. In many parts of the world, the PSTN service provider *is* an official department of the government, known as postal telegraph and telephone (PTT) entity. For example, the use of identity is regulated on the PSTN. In many countries, the PSTN operator is required by law to provide a mechanism by which a caller can prevent the disclosure of his calling party telephone number to the called party, often by dialing a star code prior to dialing the called number. Certain classes of service must be signaled over the PSTN, in addition to the calling party number; for example, in the United States, telephone calls originating from prison pay phones are specially marked in the call signaling over NNI interfaces.

The prevention of unwanted calls has also been shown to be effectively done in the PSTN using legislation. In the United States, telephone subscribers can register with the National Do Not Call Registry [1] and request that the PSTN block unsolicited telephone marketing calls. Businesses that disregard this regulation may receive heavy fines.

12.3 Gateway Security

While a PSTN gateway is essentially a user agent (or a collection of user agents) in a SIP VoIP architecture, there are some unique aspects to its operation, especially with respect to security. These include:

- Call usage by a PSTN gateway can generate significant usage and access charges.

- PSTN gateways can serve as points of attack into PSTN Telephone Switches and the Signaling System #7 (SS7) network.

- Handling of calling line ID and other identifiers in the PSTN is carefully regulated. In many jurisdictions, PSTN gateway operators that misuse or mishandle CLID could be fined and penalized.

- Telephone hackers and phone phreakers may utilize gateways from the PSTN as an attack vector into a VoIP network.

12.3.1 Gateway Security Architecture

Many of the principles of securing VoIP endpoints in general apply to PSTN gateways. PSTN gateways usually are not operated as a standalone element but usually have a proxy server in front of them. The PSTN gateway usually does not directly authenticate VoIP users. Typically, a screening proxy server is used to challenge and authenticate callers. The SIP Proxy and PSTN gateway are part of the same trust domain and maintain a secure signaling relationship that detects and rejects any messages sent to try to bypass the screening proxy as shown in Figure 12.2.

An operator of a PSTN Gateway service must authenticate users of the service so that PSTN charges can be billed to the appropriate user. However, there are some special cases that require alertness in gateway operators, such as:

- Special toll numbers (900 and 976 in the United States) often have extremely high per minute charges. Many PSTN gateway operators choose to block service to these numbers to avoid billing/charging headaches that can result. In practice, if service to such numbers is offered, PSTN gateway operators should audit calls carefully and thresholds should be set to detect unusual call behavior and throttle excessive call placement to special toll numbers.

- Calls to certain international numbers result in exhorbitant toll charges. For example, some Caribbean Islands have numbers that look

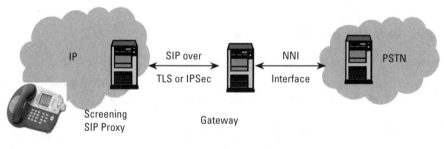

Figure 12.2 Gateway security architecture.

like "normal" North American phone numbers (1+10 digits), but access charges associated with call placement to such numbers are extremely high. PSTN gateway operators should consider screening or blocking of these numbers. As with special toll numbers, PSTN gateway operators should audit international calls carefully and set thresholds should be set to detect unusual call behavior and throttle excessive call placement.

Most PSTN operators employ sophisticated, real-time, fraud detection systems that monitor call detail records and report anomalies. A VoIP system can also utilize such systems to identify calling pattern anomalies. For example, unusually excessive call placement by a normally moderate VoIP user may indicate that the user's credentials have been stolen.

Fraud detection is not limited to PSTN gateways. Any VoIP aware device in the path can perform antifraud monitoring, generate alarms. Inline VoIP-aware security systems, like their intrusion prevention system counterparts on data networks, can (temporarily) block service when fraud is suspected. For example, a VoIP aware firewall or router could detect excessive call placement to a special toll number and block subsequent attempts by the calling party. Misuse detection can also be a passive (offline) activity: a log analysis application can inspect events from SIP proxy servers logs and identify anomalies in calling behavior across an entire organization or subscriber base.

PSTN telephone network elements have long been targets for attackers, primarily due to the centralized nature of the network itself ("attack the nucleus"). The transitive trust security relationships among telecommunications carriers makes attacks against network elements even more attractive because of the potential for escalating an attack beyond the initial target. Again, the PSTN relies heavily on physical security to protect its switching fabric. Network operations centers, remote electronics, and central office switches. are commonly protected against unauthorized entry. Physical antitampering measures protect physical cabling (the fiber plant) and equipment cabinets.

PSTN physical security policies are as rigorous as any perhaps excepting the financial industry. These policies are in place to prevent intrusion, abuse, misconfiguration, and the like.

12.3.2 Gateway Types

VoIP Gateways fall into two type, based on the level of trust they establish with the PSTN. *Enterprise gateways* are considered external network elements by telecommunications providers. They do not operate within the trust domain of any telecommunications provider and trust is not extended to them by the PSTN.

As a result, enterprise gateways connect to the PSTN using UNI protocols. Identity information passed from an enterprise gateway to the PSTN is subjected to the same screening and validation by the PSTN service provider as any subscriber identity. A *network gateway* is considered a trusted element. It is essentially considered part of the PSTN switch network and connects using NNI protocols. Validation and screening of identity information must be done by the network gateway.

These different types of gateways operate under different security policies and are subjected to different attacks. An attack on an enterprise gateway is equivalent to an attack on a UNI interface to the PSTN, and therefore, it may be possible to disrupt or steal service from a single user or the entire enterprise served by an enterprise gateway. Since the enterprise gateway is not trusted, the scope of the attack is limited to what the enterprise gateway is authorized to do by the PSTN.

An attack on a network gateway is equivalent to an attack on an NNI interface to the PSTN. Since the network gateway is a trusted element, a successful attack will give an attacker more opportunities to escalate the attack, beyond this single network element and potentially, beyond the telecommunications carrier victimized by the initial attack. The attack can expand to include any PSTN element that trusts and interconnects with this gateway. Since a trusted element in the PSTN makes identity assertions, the attacker can make a whole range of spoofed identity assertions, limited only by the geographic range of the interconnected network. (That is, a compromised network gateway interconnected with the U. S. telephone network could make false identity assertions of virtually any U.S. telephone number, but it could not assert international phone numbers as identities.)

12.3.3 Gateways and Caller ID

Identity as presented to end users in the PSTN is known as caller ID or calling line ID (CLID). Caller ID is an important service in the PSTN, used for screening, logging, and other service. Note that Caller ID is not used for billing—the PSTN has another identity mechanism known as Automatic Number Identification (ANI) which is never exposed to end users and is used for billing.

Caller ID identity assertions in the global PSTN telephone network rely on transitive trust. An identity asserted by a trusted network element, as for example, a VoIP gateway connected as a network gateway, will be trusted by other network elements in the PSTN, regardless of the actual source of the identity assertion.

Some enterprise gateways today are incorrectly interconnected with the PSTN as a network gateway, allowing incorrect identity information to leak into the PSTN.

In cases where the identity has been questionably authenticated or is not verifiable, passing no caller identity is preferable to passing potentially incorrect values. A network gateway must not pass an identity (assertion) to other PSTN network elements unless it has been properly validated. An operator who is not sure if the gateway is trusted or not should assume that the gateway is not trusted and not pass identity information other PSTN network elements.

VoIP gateways illustrate the inherent weaknesses of transitive trust and perimeter security, as described in earlier chapters. But the CLID-based identity system used by the PSTN is proving to be easily exploited. Caller ID spoofing, or "orange boxing" is actively pursued by the hacking community, and at least one obvious, albeit disreputable, commercial application exists. It helps telemarketing companies hide their identity from parties they call by displaying a number and identity that is either benign or trusted. How serious is this threat? A Google search of "fake caller id" returns a number of potential service offerings and numerous FAQs and HowTos.

12.3.4 Caller ID and Privacy

PSTN callers in many areas offer the option of blocking their caller identity. However, in the PSTN, the blocked identity is still available to other telephone switches within the trust domain. The information is conveyed over NNI interfaces but removed before sending over a UNI interface, as the trust domain is crossed. If a network gateway receives such identity information over an NNI interface, it must not pass this information to elements within the VoIP network over a UNI. A possible exception to this is if the VoIP network extends the PSTN trust domain into the IP space. If this is the case, then the border element in the IP space that enforces the trust boundary must respect this PSTN identity information, effectively making the VoIP connection an NNI interface.

Typically, a PSTN gateway will be part of a trust domain, as defined in Chapter 11 and will utilize network asserted identity techniques.

12.3.5 Gateway Decomposition

If a gateway is decomposed using the master/slave decomposition protocols described in Chapter 3, the appropriate security mechanism for those protocols must be used, for example, using IPSec to secure the MG to MGC connection. The signaling component of a gateway can also be decomposed into a signaling gateway (SG) as shown in Figure 12.3. A SG can be used to transport PSTN signaling protocols such as SS7 over an IP network using a framework known as signaling transport (SIGTRAN) [2]. This framework decomposes the various layers in the SS7 protocol.

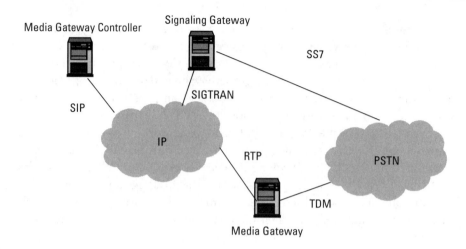

Figure 12.3 Gateway decomposition into MG, MGC, and SG.

12.3.6 SIP/ISUP Interworking

The signaling function of a PSTN gateway bridges the IP-based and PSTN networks. Gateways use detailed mapping between the two voice signaling protocols to accomplish this. For example, a SIP to PSTN gateway will follow the SIP/ISUP mapping specification [3], and interworking between H.323 and the PSTN will be performed as defined in H.246 [4]. Gateways must be tested to ensure that unusual, malformed, or invalid SIP messages do not propagate into the PSTN network. Special care must be applied to general signaling SIP messages such as INFO [5], which will be described in the next section on ISUP tunneling.

Additional security requirements are imposed on PSTN gateways that support ISUP tunneling [6], in which ISUP messages are encapsulated and carried as MIME attachments. SIP-T is a framework a mechanism that defines how PSTN gateways can pass signaling across an IP network using SIP with complete ISUP transparency, such that when a call originates from one PSTN network, is routed as SIP and RTP over an IP network, and terminated at a UA connected to a second PSTN network, no ISUP parameters or values are lost along the signaling path.

ISUP tunneling between PSTN gateways is depicted in Figure 12.4. From this architecture, it can be seen that the trust domain of the PSTN is now extended across the IP network. As such, the strongest security measures must be used including:

- Strong end-to-end and hop-by-hop authentication of gateways;

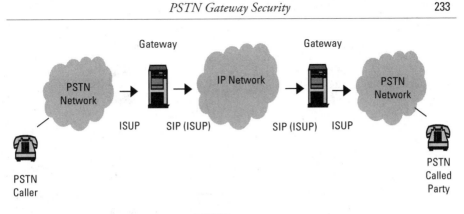

Figure 12.4 SIP-T Tunneling between PSTN Gateways.

- Encryption for confidentiality;
- Integrity protection;

Operating an ISUP tunneling gateway network without these safeguards undermines the PSTN's perimeter security.

ISUP messages are usually transported over the PSTN's tightly controlled SS7 network. VoIP interworking with ISUP must be managed and controlled very carefully or else significant exploits are possible into the PSTN network. For example, bursts of ISUP messages sent over SIP could be used to tie up all the ports on a PSTN switch, ISUP messages could also be used to launch a DoS attack on the PSTN SS7 network, and faked ISUP messages could be used to generate fraudulent telephone calls and extraneous traffic in the PSTN.

To secure against these attacks, network gateways must only accept authenticated and integrity protected SIP traffic.

12.4 Telephone Number Mapping in the DNS

ENUM [7] utilizes the Internet domain name service (DNS) to map a PSTN telephone number to an Internet address. ENUM promises to help interconnect "islands" of VoIP connectivity. For example, if a VoIP user with one service provider places a call to a VoIP user with a different VoIP provider, today, that call would be routed through the PSTN. Both VoIP service providers would utilize gateway ports and perhaps pay PSTN access and termination charges in the process. This is shown in Figure 12.5.

Using ENUM, the originating VoIP service provider would query DNS using the dialed telephone number, discover that this telephone number relates to a VoIP endpoint, and route the call over the Internet instead. This is shown in Figure 12.6.

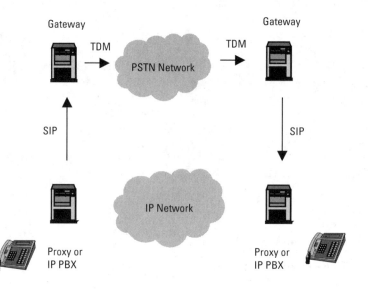

Gateway   Gateway

TDM   PSTN Network   TDM

SIP   SIP

IP Network

Proxy or   Proxy or
IP PBX   IP PBX

Figure 12.5 VoIP Islands use PSTN for interconnection.

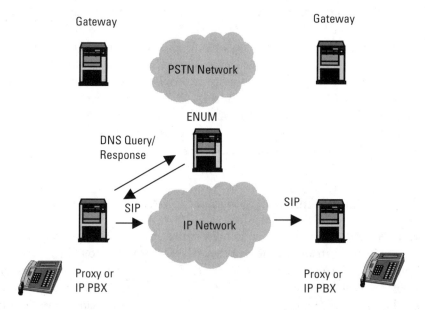

Gateway   Gateway

PSTN Network

ENUM

DNS Query/
Response

SIP   SIP

IP Network

Proxy or   Proxy or
IP PBX   IP PBX

Figure 12.6 VoIP Islands use ENUM and the Internet for interconnection.

While ENUM provides the discovery and routing mechanism to enable this scenario, service providers must be willing to exchange VoIP traffic with each other as well. This interconnection of VoIP service providers, called peering or a VoIP federation creates certain requirements for correct operation, such as:

- The ability to know the identity of the source of the calls;
- The ability to securely exchange signaling;
- The ability to securely exchange media.

When such requirements are satisfied, ENUM is used in the following manner. Starting with a telephone number in the E.164 international numbering plan format, a domain name is constructed using the following rules:

1. Invert the order of the digits
2. Insert dots '.' between each digit
3. Add the top level domain 'e164.arpa'

For example, the number 212-555-1212 in the North American Numbering Plan (NANP) in E.164 notation is +1 212 555 1212. Following the three steps above, the domain name obtained is:

```
2.1.2.1.5.5.5.2.1.2.1.e164.arpa
```

A query using the dynamic delegation discovery (DDD) system [7] is launched that returns a URI. The usage of ENUM for both SIP [8] and H.323 [9] has been defined. An example SIP AOR used in a ENUM record is as follows:

```
sip:2125551212@service-provider.example.com
```

An example H.323 URL is shown below:

```
h323:2125551212@gatekeeper.example.com
```

A key requirement for privacy is that an ENUM entry must/may not reveal any more information than the minimum amount of information. For example, the above example SIP URI does not provide anymore information than the fact that this telephone number is hosted by a particular service provider or enterprise.

Since ENUM utilizes DNS technology, ENUM is only as secure as the underlying DNS system. With ENUM, the telephone number to URI mapping provided can not be easily validated in another way, hence, the need for enhanced DNS security. ENUM is likely to be an early adopter of DNSSec, discussed in Chapter 2.

References

[1]   The U.S. National Do Not Call Registry, http://www.donotcall.gov.

[2]   Ong, L., I. Rytina, M. Garcia, H. Schwarzbauer, L. Coene, H. Lin, I. Juhasz, M. Holdrege, and C. Sharp, "Framework and Architecture for Signaling Transport," RFC 2719, October 1999.

[3]   Camarillo, G., A.B. Roach, J. Peterson, and L. Ong, "Integrated Services Digital Network (ISDN) User Part (ISUP) to Session Initiation Protocol (SIP) Mapping," RFC 3398, December 2002.

[4]   "Interworking of H.Series Multimedia Terminals with H.Series Multimedia Terminals and Voice/Voiceband Terminals on GSTN and ISDN," ITU-T Recommendation H.246, January 1998.

[5]   Donovan, S., "The SIP INFO Method," RFC 2976, October 2000.

[6]   Vemuri, A., and J. Peterson, "Session Initiation Protocol for Telephones (SIP-T): Context and Architectures," RFC 3372, September 2002.

[7]   Falstrom, P., and M. Mealling, "The E.164 to Uniform Resource Identifiers (URI) Dynamic Delegation Discovery System (DDDS) Application (ENUM)," RFC 3761, April 2004.

[8]   Peterson, J., "Enumservice Registration for Session Initiation Protocol (SIP) Addresses-of-Record," RFC 3764, April 2004.

[9]   Levin, O., "Telephone Number Mapping (ENUM) Service Registration for H.323," RFC 3762, April 2004.

13

Spam and Spit

13.1 Introduction

Unsolicited, untraceable commercial e-mail is so prevalent on the Internet that spam now exceeds legitimate e-mail by a factor of more than four to one. The value proposition for spammers is simple. For an incredibly small fraction of the cost of any other advertising and delivery method, a "merchant" can reach thousands if not millions of potential customers. A very small percentage of responses resulting in sales makes a spamming venture profitable. The "take rate" is, again, an incredibly small fraction of the response rate advertisers must obtain using bulk rate postal services.

Spammers don't have to work hard or spend money to acquire addresses for potential customers. While real-world merchants pay for mailing lists, spammers employ spambots to harvest lists of e-mail addresses from web pages, worms to steal them from e-mail address books discovered on infected PCs, or merely generate lists by prepending likely names to registered domain names in popular top level domains. Spammers don't have to spend a lot of money sending spam, either. Spam is commonly sent via PCs and servers that have been compromised through viruses or Trojans.

Once an attacker compromises a PC or server and gains administrative privileges, he can install a Simple Mail Transport Protocol (SMTP) mailer begin sending e-mails on behalf of the merchant who's hired him. (We use the term "merchant" loosely. The overwhelming majority of products and services advertised using spam are scams.) Spammers also benefit from lax administration of Internet mail servers, especially those that run as open relays. Open mail relays are mail servers that do not employ authentication for use, and they are readily exploited by spammers to hide their identity, since the spam forwarded through such relays will appear to originate from the open relay mail host rather than the spammer's mailer.

Unfortunately, some of the characteristics of e-mail have some applicability to VoIP. The cost of sending e-mail is essentially zero, and the ability to send a message to thousands and thousands of recipients trivial. In many deployments, the cost of placing VoIP calls over the public Internet also has a similar zero cost. Sending a request to ring multiple VoIP destinations at the same time, known as "forking" is also relatively easy to do and inexpensive. Even where VoIP toll charges may be incurred, parties who place unsolicited voice calls using VoIP can evade the charges by originating the calls from a UA installed on a compromised PC or server, as spammers do with e-mail today.

Unless we learn from our experiences with spam e-mail, VoIP spam, or spit (spam over internet telephony) will quickly grow out of hand. In this chapter, we will look at some of the technical challenges in dealing with spam and compare them to possible techniques useful to fight spit. Note that there is another entire aspect of spam that is sociological in nature, and requires nontechnical means to combat. These issues and approaches will not be discussed here.

13.2   Is VoIP Spam Inevitable?

Spam exploits deployment, implementation, and technical characteristics of Internet e-mail:

1. Since the 1980s, Internet e-mail is largely exchanged frequently over the Internet among mail systems hosted by ISPs, colleges and universities, private organizations, government agencies, and even by individuals who host their own mail servers. E-mail is one of the Internet's "killer" applications and will remain so.

2. The cost of sending bulk e-mail is essentially zero. As we explained in the introduction to spam, although there are costs associated with the bandwidth and connection time to the Internet, those who spam typically do not bear the cost.

3. Sending the same message to large numbers of recipients is trivially performed, with little incremental cost to the sender over sending a single e-mail.

4. There is no widely used mechanism for asserting e-mail identity. As a result, most spam has no traceable identity. Even when traceable, the identities are often spoofed. In addition, there are no widely deployed

mechanisms to prove an individual sent a spam e-mail in a non-repudiatable manner.

5. Spam has no easily traceable return e-mail path. In some cases, the return path is obscured or erased, while in others, it leads to spoofed senders and compromised systems.

Considering these characteristics in the context of VoIP, we can make the following observations:

1. Today, few VoIP providers, enterprises, and users freely exchange VoIP traffic over the Internet. We have every reason to believe, however, that VoIP adoption will grow. Like e-mail, VoIP has the potential to be one of Internet's killer applications. It follows that spit will become economically attractive.

2. The cost of placing VoIP calls in volume will be free (or essentially so for spitters). Unless security practices improve measurably, it is likely that spitters will exploit poorly secured VoIP systems to send spit.

3. Spitters have several methods for placing the same VoIP call from which multiple recipients can choose. For example, using an approach known as "forking" [1], it is easy to send a SIP request to multiple destinations at the same time. Other methods are under development as well. The IETF SIP Working Group is standardizing exploders [2], B2BUAs designed to send a single request to multiple destinations. A UA can use an XML encoded list of destination URIs to send a single call request, and have a copy delivered to every destination in this resource list [3]. The IETF standards are being written to require the opt-in consent of all users in the resource list, known as the consent framework [4]. However, it is likely that this approach, once standardized, will be utilized without these safety features.

   The absence of an asserted identity is unfortunately as true for VoIP systems today as it is for email senders. The `From` header field in SIP request messages is set by the originating user agent. The `P-Asserted-Identity` [5] header field discussed in Chapter 11 is only applicable within a trust domain, and does not solve the general interdomain identity problem. The enhancements for SIP identity [6] solves this problem in an elegant way, however it is not yet deployed.

4. SIP does not have a verifiable return path. Separate mechanisms are used to originate versus received SIP requests.

At this time, we reluctantly concede the inevitability of spam for VoIP.

13.3   Technical Approaches to Combat E-Mail Spam

A number of technical approaches are used to combat spam in e-mail today [7]. Many have been developed within the Internet Research Task Force (IRTF) Anti-Spam Research Group (ASRG) [8]. Others have been developed by large e-mail service providers and commercial software and hardware security vendors.

Some common antispam measures include:

- Filtering by sender identity (black and white listing);
- Grey listing;
- Challenge/response mechanisms;
- Content filtering in servers and PCs;
- Reputation services.

These approaches will be discussed in the following sections.

13.3.1 Filtering Spam Using Identity Information

Client side-filtering, the processing by a user's mail client, or preprocessing performed on behalf of clients by an antispam server application based on sender identity, is commonly practiced today. E-mail is approved (white listed) or banned (black listed) based on originating e-mail address. This form of filtering is difficult in e-mail since there is no way to verify identity in e-mail. Spammers routinely spoof e-mail senders, so filtering can result in misclassification. Signing e-mail messages with digital signatures (S/MIME and PGP) that relies on digital certificates, a PKI, or directory infrastructure to publish public keys would measurably improve filtering, but the adoption rate of these and other "lightweight" authentication schemes for signing e-mail remains extremely small relative to the e-mail user population.

SMTP message transit agents (MTAs) can perform server-side filtering on the IP address of the MTA that established a TCP connection to process mail. MTAs can perform a reverse lookup (in-addr.arpa) on the IP address to see if a domain name can be determined. Many spammers send e-mail from hosts whose IP addresses do not have a reverse (PTR) record in the DNS. If an MTA discovers a negative response to a reverse lookup, it can as evidence that the sending MTA is not trustworthy and that messages it wants to send may be spam. Another approach is to try to deduce whether the MTA's IP address is associated with DHCP pools of dynamic addresses. SMTP operators can also refer to lists of IP addresses of known spamming hosts and open SMTP relays and block any incoming connections; alternatively, SMTP mailers can maintain a list of trusted SMTP hosts and relays and only accept traffic from members of this list.

Spam e-mail is usually sent from a different set of SMTP relays than those used to send most of the messages from that domain. The DomainKeys [9] and SenderID [10] antispam solutions attempt to contain spam by providing DNS records for trusted mail relays. When an e-mail message is received, the MTA processing the message issues a DNS query on the domain name in the From address field to determine if this message is from an authorized SMTP relay; a negative response from the DNS indicates that the sending MTA is a spamming host. These approaches suffer from the requirement that every e-mail operator must enter the appropriate records into the DNS, and that all operators implement the mechansim and reject sent mail that fails this check.

Another approach borrows from the hacking community's technique for OS fingerprinting [11] and looks for certain protocol composition and behavior anomalies. Analysis of how commercial and popular OS implementations compose and respond to correct and malformed TCP and IP packets reveals that these implementations are sufficiently unique as to be distinguishable. The uniqueness in responses yields the equivalent of a fingerprint. Similar analysis can be used to fingerprint and blacklist a spamming server.

13.3.2 Grey Listing

"Grey listing" [12] is a white listing sender identity approach with a twist. The entries in the white list have three fields: a sender e-mail address (S), a recipient email address (R), and the IP address (I) of the e-mail server that requested that a message from S to R. This list is not preconfigured, but rather, built over time, as follows. When an e-mail server receives an e-mail message, it composes a triplet (S, R, I). The server compares this triplet against its white list (which is initially empty). If the triplet is one that the mail server hasn't seen before, it responds to I by temporarily rejecting the e-mail message (SMTP defines a 450 error message for just this purpose).

Most spamware are barebones implementations, designed to send spam messages as fast as possible and ignore error messages, so they do not retry bounced e-mail. Mail server implementations that conform to the SMTP specifications will wait for a while, and then attempt to resend the message. A grey listing e-mail server thus concludes that the triplet (S, R, I) is a legitimate e-mail exchange based on receipt of a retransmitted message. The triplet is added to the white list. Grey listing requires little configuration, maintenance, and processing overhead.

Spammers may try to outfox grey listing by altering the way they react when they receive temporary reject messages, but thus far, they have not succeeded.

13.3.3 Challenge/Response (Sender Verification)

Challenge/response mechanisms treat all unknown senders as potential spammers, and challenge the sender to prove its identity. Only after a satisfactory response has been received is the sender whitelisted. The challenge is chosen to be easy for humans but difficult for computers, such as identifying letters or numbers in a graphic, called the Completely Automated Public Turing Test to Tell Computers and Humans Apart or CAPTCHA [13]. CAPTCHA is useful for preventing spammers from using automated methods to create email accounts or relay spam through web forms and weblogs (blogs).

13.3.4 Distributed Checksum Filtering (DCF)

One easily observable characteristic of spam e-mail is that thousands, if not millions, of copies of a single message are forwarded to MTAs across the Internet. Individual MTAs can black or grey list a message as spam, but they do so independently. If UAs and MTAs could classify an e-mail message as spam, and quickly notify other MTAs, the resulting *distributed filtering* would be more effective. Two DCF antispam measures tackle spam in exactly this manner.

Vipul's Razor is a DCF method that combines spam reporting, a reputation system, and checksum exchange. User agents register their identities with Razor servers so they can report spam. When an MTA receives a report that an e-mail message is spam, it adds a hash computed over the spam e-mail to a database. Each time an MTA processes an arriving e-mail it computes a hash over the message and compares this to the entries in a distributed spam catalog, which can be queried by mail clients and MTAs. If the computed hash matches a Razor catalog entry, the message is spam.

Trust plays an important role in Vipul's Razor. Users contribute to spam filtering by reporting spam e-mail, so a UA's reputation, that is the degree of trust the Razor servers place on a UA's ability to classify spam, is extremely important. A UA's reputation increases or deteriorates over time, based on correct classification. Frequent misclassifications hurt a UA's reputation to discourage spammers from abusing the system.

Distributed Checksum Clearinghouse (DCC) uses distributed scoring instead of spam reporting and cataloging. A participating MTA computes a hash (X) on each message it processes and maintains a counter (N) representing the number of copies of X it receives. MTAs distribute the tuplet (X, N) all MTAs participating in DCC. High hash counts are used to distinguish spam from legitimate e-mail.

DCF solutions often white list legitimate mailing lists to prevent misclassification.

13.3.5 Content Filtering

Content filtering is perhaps the most widely used antispam measure. Approaches based on scanning actual content for spam characteristics have been very successful at reducing the amount of spam that humans process. Bayesian analysis is one of the most well known and popularly used content filtering methods. Bayesian filters use statistical analysis to determine the probability that an email is spam. The basis for filtering is observation and probability. By analyzing large samplings of spam (and there are certainly enough samples), researchers observed that certain words, word patterns, and the frequent use of key words in an e-mail message provide strong indications that a message is spam. It follows that spam probability can be calculated analyzing the contents (and mail headers) of incoming e-mail messages against a database of words and word sets identified as spam. The database can be seeded with known words and patterns, or users can train their e-mail clients based on manual classification of e-mail as spam [14, 15].

13.3.6 Summary of Antispam Approaches

Spam remains an enormous problem for e-mail users. Researcher in academia and the commercial world continue to introduce new techniques for combating spam. Future antispam measures are likely to focus on making spam less profitable. Many experts conclude that reducing spam volume is a function of economics, and that if the expense of generating spam is increased and the profits derived from spamming decrease, the problem will disappear.

Table 13.1 summarizes and compares the sample of approaches we have discussed.

13.4 VoIP and Spit

Some of the antispam approaches discussed in the previous section have application to dealing with VoIP spit. These approaches will be discussed in this section along with a discussion about how many of the VoIP approaches detailed in earlier applications can also be applied.

VoIP has some useful greylisting approaches for combating spit. For example, as many VoIP endpoints can be reached by the PSTN, suspicious calls can be redirected to the PSTN, forcing the caller to use resources and pay PSTN charges, something a spammer would not do. To establish a VoIP call, the spammer needs to have a substantial interaction, requiring much more resources than just sending a message.

Many of the identity and security features described in this book are readily applicable to the spam problem. For example:

Table 13.1
Comparison of E-Mail Antispam Techniques

Antispam Technique	Characteristics	Evaluates	Comments (Issues)
White- and black-listing	List permitted or denied mail senders	Users and domains in From/To mail headers Mail server, domain, IP	Vulnerable to spoofing, requires list maintenance
Greylisting	Rejects "never before seen" e-mail Self-configuring	From IP address Sender e-mail address Recipient e-mail address	First time legitimate mail delayed along with spam May require white lists
Keyword content filtering	Scan and drop or reject email containing "bad words"	Message body and subject	Subject to misclassification
Heuristic content filtering (e.g., Bayesian)	Scan and score e-mail for spam "signatures" (e.g., statistical analysis)	Message body and headers	Signatures require frequent update
Sender verification	Sender must prove his identity with non-repudiable signature	User input (Visual captcha) Special headers (Haiku) Tag reply address (TMDA)	Universal adoption or user action required
Distributed collaborative filtering	MTAs checksum every message processed	Use checksum count as indication that message is spam (e.g., Vipul's Razor)	Universal adoption required; subject to misclassification

- SIP identity enhancements provides a cryptographically assured identity that makes whitelisting possible.

- TLS proxy to proxy exchanges allows certificate checking and rejecting anonymous connections or ones that do not have a valid certificate.

- VoIP may utilize clearinghouse or federation models that make it more difficult for spammers to inject messages.

- Secured media sessions makes it impossible for "media spamming" in which the spammer monitors the signaling to detect calls then tries to inject spam RTP packets.

However, there are some problems in using conventional security mechanisms to defend against spit, such as:

- Perimeter B2BUA SIP routing approaches make end-to-end identity much more difficult and may help cover the tracks of spammers.

- B2BUAs will be a prime target for hackers—due to their trusted nature, they can be a perfect source of spam and spit if compromised.

- SIP can carry arbitrary message bodies and content and could be used as a firewall transport mechanism to penetrate perimeter security and spread viruses.

Finally, SIP has an existing amplification attack that could be exploited. A SIP client receiving an INVITE will begin sending RTP packets to the IP address in the SDP as soon as the user answers. The IP addresses in the SDP need not correspond to the IP addresses in the SIP signaling path. In fact, in the gateway decomposition models described in Chapter 3, the IP addresses will always be different. As such, an attacker could send an INVITE and put the target's IP address in the SDP and send to an unwitting third party. If the resulting media session is high bandwidth, and similar INVITEs are sent to other third parties, the result could be a DDoS, as described in Chapter 4. Note that the target of the DDoS does not even need to be a VoIP or SIP device—as all it sees is a flood of packets.

One way to protect against becoming an unintentional participant in this type of DDoS is to utilize the Interactive Connectivity Establishment (ICE) protocol [16]. Before sending RTP packets, ICE sends test STUN [17] packets and will only send RTP to IP addresses and ports that it has received a corresponding STUN packet from the other party.

Perhaps the best approach against spit is to combine a strong cryptographic identity with a reputation service. The reputation service can act as a certificate authority, issuing certificates to domains who agree to abide by a set of reasonable use rules that would rule out spit. Any spit within this system would be traceable to the originating domain who would be held responsible.

VoIP systems should be selectively and carefully opened up, with authenticated signaling, preferably using TLS. This will provide minimal protection against spit.

13.5 Summary

This chapter has discussed the technical origins of e-mail spam and some of the technical approaches used to combat it. The similarities and differences between e-mail spam and VoIP spit were then discussed, with some potential solutions presented. However, spam is a sociological problem, and any complete solution will likely also be sociological and include legislation and peer pressure against both the spammer and those who inexplicably patronize the businesses who utilize these methods.

References

[1]   Rosenberg, J., H. Schulzrinne, G. Camarillo, A. Johnston, J. Peterson, R. Sparks, M. Handley, and E. Schooler, "SIP: Session Initiation Protocol," RFC 3261, June 2002.

[2]   Camarillo, G., and A. Roach, "Framework and Security Considerations for Session Intiation Protocol (SIP) Uniform Resource Identifier (URI)-List Services," IETF Internet-Draft, work in progress, April 2005.

[3]   Roach, A., J. Rosenberg, and B. Campbell, "A Session Initiation Protocol (SIP) Event Notification Extension for Resource Lists," Work in progress, January 2005.

[4]   Rosenberg, J., G. Camarillo, and D. Willis, "A Framework for Consent-Based Communications in the Session Initiation Protocol (SIP)," IETF Internet-Draft, work in progress, July 2005.

[5]   Jennings, C., J. Peterson, and M. Watson, "Private Extensions to the Session Initiation Protocol (SIP) for Asserted Identity Within Trusted Networks," RFC 3325, November 2002.

[6]   Peterson, J., "Enhancements for Authenticated Identity Management in the Session Initiation Protocol (SIP)," draft-ietf-sip-identity-03, work in progress, September 2004.

[7]   "Technical Responses to Spam," Taughannock Networks, http://www.taugh.com/spam tech.pdf, November 2003.

[8]   Internet Engineering Research Task Force (IRTF) Anti-Spam Research Group (ASRG), http://asrg.sp.am

[9]   Yahoo's DomainKeys http://antispam.yahoo.com/domainkeys.

[10] Microsoft's Sender ID Framework, http://www.microsoft.com/senderid/.

[11] Fyodor, "Remote OS Detection via TCP/IP Fingerprinting," http://www.insecure.org/nmap/nmap-fingerprinting-article.html.

[12] Harris, E., "The Next Step in the Spam Control War: Greylisting," http://projects.puremagic.com/greylisting/whitepaper.html.

[13] The CAPTCHA Project, http://www.captcha.net.

[14] Haskins, R., and D. Nielsen, *Slamming Spam*, Reading, MA: Addison-Wesley, 2004.

[15] Graham, P., "A Plan for Spam," http://www.paulgraham.com/spam.html.

[16] Rosenberg, J., "Interactive Connectivity Establishment (ICE): A Methodology for Network Address Translator (NAT) Traversal for Multimedia Session Establishment Protocols," IETF Internet Draft, October 2005.

[17] Rosenberg, J., J. Weinberger, C. Huitema, and R. Mahy, "STUN—Simple Traversal of User Datagram Protocol (UDP) Through Network Address Translators (NATs)," RFC 3489, March 2003.

14

Conclusions

14.1 Summary

Voice is an exciting and emerging application for the Internet. The suite of applications that enable Voice over IP are even more intriguing, because they are not simply applications in and of themselves, but enabling technologies for a much richer set of real-time applications. Business users and consumers have enthusiastically adopted simple text messaging, as well as image and video exchanges that are offered over cellular phone systems. It thus appears safe to predict that with more bandwidth, memory and faster processors, VoIP handheld devices will allow us to experience much better Internet use from PCs and laptops not imagined possible even a short while ago.

What is troubling, and what motivated the production of this book, is how important security is to voice applications, how differently society perceives voice security, and how easy it will be to deploy VoIP incorrectly, with potentially disastrous results. Rational or not, most individuals and businesses place a higher premium on the availability of phone service and the confidentiality of telephone conversations than they do Internet communications. In most countries, eavesdropping on telephone conversations and "cloning phone numbers" is illegal, and telephone system operators have invested considerable capital and expertise to comply with laws and regulations concerning wiretapping, identity fraud, and other abuses of telephony deemed unlawful. Similar laws have recently been enacted to protect e-mail and other forms of Internet correspondence, but we continue to struggle to adopt technology to prevent unlawful activities, making enforcement and prosecution nearly impossible.

Many people falsely conclude that laws currently protecting PSTN-originated and delivered phone calls uniformly apply to VoIP, and that VoIP operators have the technology and wherewithal to protect "the phone network." We hope we have corrected this false assumption, by providing a sufficient number of sobering examples of VoIP misuse to convince you to implement the broadest security measures possible when you adopt VoIP.

As you attempt to tackle VoIP security, we recommend that you consider the following issues carefully.

14.2   VoIP Is Still New

Within the context of Internet security, new is rarely considered "a good thing." New applications and protocols are historically green fields for attackers, and this clearly holds true for VoIP. Testing VoIP applications and protocol implementations to isolate and correct "bugs" is a significant issue in VoIP deployment. Expanding industry best practices for securing data networks to encompass VoIP is a significant undertaking as well. Testing and auditing VoIP systems and networks should thus be given ample consideration and funding as VoIP is deployed. Finally, it is important to recognize that while we have attempted to identify as many threats against VoIP systems as is practical, it is unlikely that we have yet seen the full spectrum of traditional voice and data attacks aimed against VoIP and, therefore, can only speculate that many more innovative, and voice-specific attacks are forthcoming. In designing VoIP security measures for your organization, it will be necessary to keep informed of intrusion detection systems that are able to identify "never before seen" activities and react according to your policy input.

14.3   VoIP Endpoints Are New

VoIP network designers must also consider that user adoption and anticipated life cycle of mobile VoIP endpoint devices will rival cellular phones. This complicates the already challenging issue of securing network endpoints. Not only will new endpoint devices (CPU, bandwidth and memory challenged, with lightweight OSs that are not easily hardened) create new attack opportunities, but the time window to design and implement security measures versus the time individuals will make use of the current handheld devices before they seek new and improved devices is disturbingly small. It is important to study endpoint devices and choose devices that can be configured securely or accommodate third-party software security. Investigate vendors carefully, and choose manufacturers with good security track records (e.g., whose products and software appear *infrequently* on bug reporting lists). Identify target requirements for a secure VoIP endpoint for your organization (or personal use!). Requirements for a secure VoIP endpoint are presented in Table 14.1.

14.4 VoIP Standards Are Not Complete

Like most of the Internet protocols and applications that preceded VoIP, security is largely an afterthought. First generation products do not provide a rich set of security services, and we are arguably 9–18 months away from seeing many of the emerging security solutions and standards described in this book in commercial product forms. We recommend that you develop a multiyear plan for secure VoIP deployment that makes best use of security measures and countermeasures at hand today, and establishes a staged introduction of improved security measures as they are adopted as standards and appear in commercial product.

14.5 Base VoIP Security on Best Current Security Practices for Data

Many best practices for protecting data networks are appropriate for VoIP. We obviously advocate layered security measures. Begin with an authentication framework that provides you with the strongest practical solution for asserting identity, preferably one that provides non-repudiation and is effective in defeating impersonation attacks. Follow industry best practices for securing endpoint devices and servers. Investigate endpoint and admission control security measures and adopt secure tunneling at least for signaling if not for signaling and media as well. Identify target requirements for secure VoIP servers for your organization. Table 14.2 provides an example set of requirements.

Develop a corporate "mental picture" of a secure VoIP session for use as the basis for security policy definition. One possible characterization of a secure VoIP session is illustrated in Figure 14.1 and summarized here:

Table 14.1

Protocol Requirements for Secure SIP VoIP Endpoint

Protocol	Specification	Details
SIPS	RFC 3261	Secure SIP with TLS support
SRTP	RFC 3711	Secure RTP
Digest	RFC 2717	HTTP Digest Authentication
SIP Certs	*	SIP certificate service for retrieval of credentials
Connection Reuse	*	Reuse of TLS connection after mutual auth

Table 14.2
Protocol Requirements for Secure SIP VoIP Servers

Protocol	Specification	Details
SIPS	RFC 3261	Secure SIP with TLS support
Digest	RFC 2717	HTTP Digest Authentication
SIP Identity	*	SIP enhanced identity mechanism
SIP Certificates	*	SIP certificate service for retrieval of certificates and credentials
Connection Reuse		Reuse of TLS connection after mutual auth

- Registration and mutual authentication between UA and SIP server;
- Retrieval of credentials (private key) using SIP certificate service;
- Retrieval of certificate of called party using SIP certificate service;
- Session negotiation over a secure connection with enhanced identity to exchange session keys for media session;
- Validation of identity of calling party;
- Retrieval of certificate of calling party using SIP certificate service;
- Establishment of a secure media session.

14.6   VoIP Is a QoS-Sensitive Data Application

This admittedly introduces some seemingly unique problems for network designers, especially for networks that have not investigated QoS before. These problems are solvable, but in many cases require more expertise and greater cooperation among IT, IS, and voice departments.

14.7   Merging Public and Private VoIP Services Will Be Problematic

We only need to look at the difficulties enterprises experience with public instant messaging services to appreciate how quickly VoIP security issues will become enterprise security priorities. Like the early versions of public IM services, public VoIP services will not offer enterprise-strength security. Software clients and consumer-priced hard phones will suffer from "haste-to-market" implementations and will be easily exploited, and the predictable reaction from IS departments worldwide will be to "block public VoIP, let's roll our own

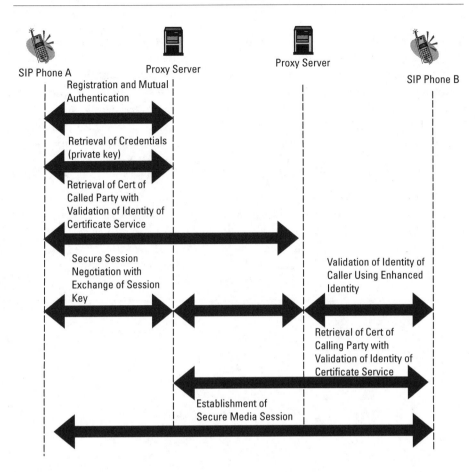

Figure 14.1 Secure VoIP session establishment high-level flows.

secure solution." A better approach may be to evaluate VoIP endpoint technology, identify devices or software that will meet business needs and can be secured, and make these the corporate standard. It should also be recognized that employees will adopt VoIP on home office networks, develop a set of best practices, and create incentives for implementing them.

Federation of enterprise VoIP and other real-time communications systems will be implemented provided adequate security and identity mechanisms are utilized.

14.8 Concluding Remarks

The authors sincerely hope and anticipate that the security of VoIP systems in particular and Internet-based communications systems in general will steadily improve over time. We feel some optimism based on the emerging work on asserted identities. As interdomain identity and secure communications are realized, business users and consumers alike should have strong confidence in their communication tools, and not suffer from out of control spam and spit. The requirements and protocols for securing SIP signaling are relatively well understood. The requirements for securing media sessions are also well understood, but we anticipate that the performance factor will cause media security to lag somewhat behind signaling, at least initially. As is the case for data today, VoIP security is not solved by technology alone. Organizations must garner a greater appreciation for both the need *and* utility of security.

We hope we have provided you with some useful insight into the VoIP security problem, and that you are able to make use of some of the recommendations herein. We welcome hearing from you. Alan can be reached at alan@sipstation.com and Dave can be reached at dave@corecom.com.

Index

For further information on these and other Artech House titles,
including previously considered out-of-print books now available through our
In-Print-Forever® (IPF®) program, contact:

Artech House
685 Canton Street
Norwood, MA 02062
Phone: 781-769-9750
Fax: 781-769-6334
e-mail: artech@artechhouse.com

Artech House
46 Gillingham Street
London SW1V 1AH UK
Phone: +44 (0)20 7596-8750
Fax: +44 (0)20 7630-0166
e-mail: artech-uk@artechhouse.com

Find us on the World Wide Web at: www.artechhouse.com